Two Manuals Combined into One Book

Military Police Operations

And

Tactical Employment of Nonlethal Weapons

Department of Defense

Pentagon
Publishing

***FM 3-19.1** (FM 19-1)

Field Manual
No. 3-19.1

Headquarters
Department of the Army
Washington, DC, 22 March 2001

Military Police Operations

Contents

Distribution Restriction: Approved for public release; distribution is unlimited.

*This publication supersedes FM 19-1, 23 May 1988.

Preface

The Military Police (MP) Corps supports the commander across the full spectrum of military operations. This manual is the foundation for all MP doctrine as it relates to this support. It communicates to all levels of leadership and staffs how the MP provide a flexible and lethal force capable of operating across this full spectrum. As the keystone manual, it identifies what the MP train on and how their forces are organized and equipped in support of all Army echelons. Additionally, this manual provides guidance that can be used by United States (US) Army service schools, students, sister services, and federal agencies.

This manual is based on the purpose, organization, responsibilities, and goals of the US Army as set forth in Field Manuals (FMs) 100-1 and 3-0, as well as corps, division, and brigade manuals. Additionally, this manual is fully compatible with current joint, multinational, and interagency doctrine.

Appendix A contains a metric conversion chart.

The proponent of this publication is HQ TRADOC. Send comments and recommendations on Department of the Army (DA) Form 2028 directly to Commandant, US Army Military Police School (USAMPS), ATTN: ATSJ-MP-TD, 401 Engineer Loop, Suite 2060, Fort Leonard Wood, Missouri 65473-8926.

Unless this publication states otherwise, masculine nouns or pronouns do not refer exclusively to men.

Chapter 1

Introduction

The MP Corps has a strong history evolving over the past five decades. We, as a corps, continue to transform our organizations and doctrine as we have in the past to support the Army in the active defense of the 1970s, the AirLand battle of the 1980s, and now the force-projection doctrine of the 1990s. Our five MP functions clearly articulate the diverse role the MP play across the full spectrum of military operations. We cannot bask in our successes, nor reflect or celebrate. Our charter is to continue our legacy of stellar performance and strive to perfect it.

BG Donald J. Ryder

When the Army developed the Active Defense strategy in 1976, the US was facing the Cold War scenario of central Europe. Military strategy and doctrine were related to a single, focused threat that revolved around the countries in the Warsaw Pact. We were an outnumbered and technically inferior force facing an armor-dominated European battlefield. The MP Corps supported the Active Defense strategy by tailoring its forces to meet the threat. In 1982, when the AirLand Battle strategy was developed, US forces were still outnumbered, but were no longer technically inferior. Still threat-based and focused on a central European conflict, the AirLand Battle strategy used a relatively fixed framework suited to the echeloned attack of soviet-style forces. It delineated and clarified the levels of war; emphasized closed, concerted operations of airpower and ground forces; balanced the offense and the defense; and highlighted the synchronization of close, deep, and rear operations. MP doctrine kept pace with the Army's AirLand Battle strategy by supporting the battlefield commander through four basic missions—battlefield circulation and control, area security (AS), enemy prisoner of war (EPW), and law and order (L&O).

OVERVIEW

1-1. In October 1983, MP capabilities in the AirLand Battle strategy were tested during operation Urgent Fury in Grenada. The MP performed missions that ranged from assisting the infantry in building-clearance operations to assisting Caribbean peacekeeping forces in restoring L&O. These actions secured the MP's place in the combat-support (CS) role, demonstrating the professional knowledge and flexibility necessary for rapid transition from combat to CS to peacetime missions. The changing battlefield conditions of operation Urgent Fury set the stage for the demand of MP units today.

1-2. Evolving simultaneously with the changing definition of the modern battlefield, MP performance in Operations Hawkeye, Just Cause, and Desert Shield/Storm galvanized their ability to perform at any point along the operational continuum. With the publication of FM 100-5 in 1993, the Army adopted the doctrine of full-dimensional operations, relying on the art of battle command to apply those principles and to shift the focus from AirLand Battle to force-projection doctrine. This new doctrine was based on recent combat experience in a multipolar world with new technological advances. Already trained and expected to perform in this new strategy, MP support was already in place and fully operational. The MP continued to perform their basic battlefield missions and to refine their capabilities while supporting the battlefield commander as he deployed to contingency operations throughout the world.

1-3. In 1996, the MP Corps went through a doctrinal review process to determine if it was properly articulating its multiple performance capabilities in support of US forces deployed worldwide (see Appendix B). The review process identified the need to restructure and expand the EPW mission to include handling US military prisoners and all dislocated civilians. This new emphasis transformed the EPW mission into the internment and resettlement (I/R) function. The review process also identified the need to shift from missions to functions. In the past, the four battlefield missions adequately described MP capabilities in a mature theater against a predictable, echeloned threat. However, that landscape is no longer valid. Accordingly, the four MP battlefield missions have become the following five MP functions:

- Maneuver and mobility support (MMS).
- AS.
- L&O.
- I/R.
- Police intelligence operations (PIO).

1-4. These new MP functions are shaped by the following factors:

- The application of stability and support operations where the integration of joint, multinational, and interagency capabilities are common occurrence.
- The lack of traditional linear battlefields, requiring theater commanders in chief (CINCs) to request forces that meet a specific function to accomplish operational requirements.
- The impact of asymmetric threats (such as drug traffickers and terrorist factions) and the effects of man-made and natural disasters.
- The impact of advances in information and communication technologies and specifically in understanding the increased vulnerabilities presented by these technologies.

1-5. Articulating MP capabilities along functional lines benefits the MP and the Army echelon commander as well as the combatant commander. Since there is a multinational, interagency, and sister-service overlap of security services, the importance of including MP leaders and staffs early in the operational planning process cannot be overemphasized. This means before units are designated, unit boundaries are drawn, and unit missions are assigned. Early involvement ensures the proper development of common security responsibilities, communication and connectivity, liaisons, processes, and the rules of interaction between all forces. The ultimate goal should be the optimal, phased employment of MP forces in support of a commander's operational plan. MP functions not only reflect and capture current capabilities, they define the MP Corps in the twenty-first century.

1-6. As the Army reshapes and focuses its resources on transformation, Force XXI, and other redesign efforts, the MP Corps stands proud and ready to support this progress and reiterate its commitment to assist, protect, and defend.

OPERATIONAL FRAMEWORK

1-7. The operational framework consists of the arrangement of friendly forces and resources in time, space, and purpose with respect to each other, the enemy, or the situation (see Figure 1-1). The operational framework for Army forces (ARFOR) rests within the combatant commander's theater organization. Each combatant commander has an assigned geographical area of responsibility (AOR), also called a theater, within which he has the authority to plan and conduct operations. Within the theater, joint-force commanders at all levels may establish subordinate operational areas such as areas of operation (AOs), joint-operations areas (JOAs) and joint rear areas

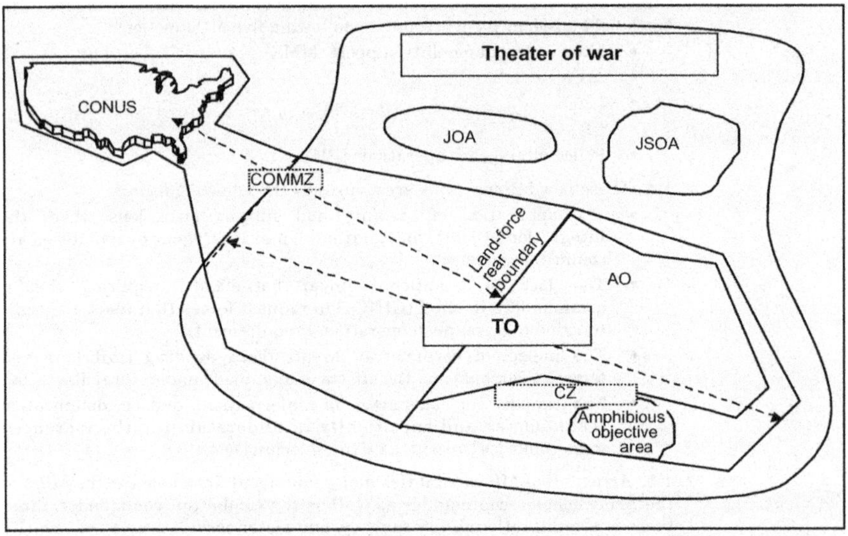

Figure 1-1. Operational Framework

(JRAs). The JRAs facilitate the protection and operation of bases, installations, and forces that support combat operations. When warranted, combatant commanders may designate theaters of war, theaters of operations (TOs), combat zones (CZs), and communications zones (COMMZs).

1-8. A theater of war is that area of air, land, or water that is, or may become, directly involved in the conduct of the war. A theater of war may contain more than one TO. It does not normally encompass the geographic combatant commander's entire AOR. A TO is a subarea (defined by a geographic combatant commander) within a theater of war in which specific combat operations are conducted or supported.

1-9. A CZ is the area required by combat forces for conducting operations. It normally extends forward from the land force's rear boundary. The COMMZ is the rear part of the TO (behind but contiguous to the CZ) that contains the lines of communication (LOC) and provides supply and evacuation support. Other agencies required for the immediate support and maintenance of field forces may also be located in the COMMZ. The COMMZ spans back to the continental US (CONUS) base, to a supporting combatant commander's AOR, or both.

1-10. An AO is an operational area defined by the joint-force commander for land and naval forces. An AO does not typically encompass the entire operational area of the joint-force command (JFC), but it should be large enough for component commanders to accomplish their mission and protect their forces. Army commanders use control measures to describe AOs and to design them to fit the situation and take advantage of the joint force's capabilities. Commanders typically subdivide the assigned AO by assigning subordinate-unit areas. These subordinate-unit areas may be contiguous or noncontiguous (see Figure 1-2). When friendly forces are contiguous, a boundary separates them. When friendly forces are noncontiguous, the concept of operations links the force's elements, but the AOs do not share a boundary. The intervening area between noncontiguous AOs remains the responsibility of the higher headquarters.

BATTLEFIELD ORGANIZATION

1-11. Battlefield organization is the arrangement of forces according to purpose, time, and space to accomplish a mission. Battlefield organization has both a purpose- and spatial-based framework. The purpose-based framework centers on decisive, shaping, and sustaining (DSS) operations. Purpose unifies all elements of the battlefield organization by providing the common focus for all actions. However, forces act in time and space to accomplish a purpose. The spatial-based framework includes close, deep, and rear areas. Despite the increasing nonlinear nature of operations, there may be situations where commanders describe DSS operations in spatial terms. Typically, linear operations involve conventional combat and concentrated maneuver forces. Ground forces share boundaries and orient against a similarly organized enemy force. In such situations, commanders direct and focus simultaneous DSS operations in deep, close, and rear areas, respectively (see FM 3-0).

1-12. The MP battlefield organization supports every Army echelon, from the Army service component command (ASCC) and the theater-support command

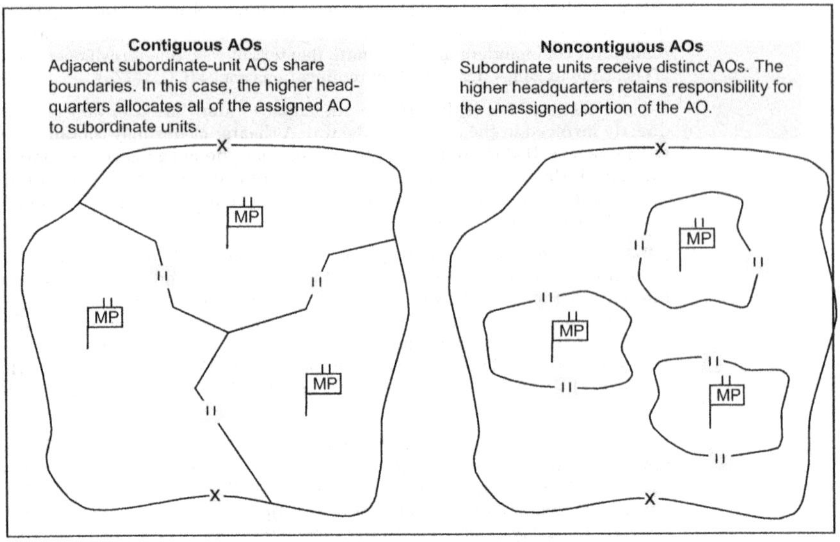

Figure 1-2. Contiguous and Noncontiguous AOs

(TSC) to the maneuver brigade. Regardless of the battlefield organization (purpose or spatial based), MP support to the Army commander is based on available resources and mission, enemy, terrain, troops, time available, and civilian considerations (METT-TC).

1-13. MP support throughout the theater of war may include MP units in the JOA and in the TO. If the combatant commander designates a COMMZ and a CZ within his TO, MP support will come from the established MP modified table of organization and equipment (MTOE) at the subordinate-command echelon. MP support to the JOA is also provided based on METT-TC and available MP assets. Typical MP support may include an I/R brigade liaison detachment (BLD), MP brigades and battalions, a division MP company, a military-working-dog (MWD) team, an L&O team, and a customs team. Figure 1-3, page 1-6, depicts a typical MP organization throughout the TO. In the COMMZ, Figure 1-3 depicts the different types of MP units that are assigned to echelons above corps (EAC) (the ASCC or the TSC). In the CZ, Figure 1-3 depicts the different types of MP units that are assigned to corps, division, and the separate brigades.

SUPPORT IN THE COMMZ

1-14. MP support in the COMMZ is provided by an array of multifunctional MP units. The following MP units provide MP support to EAC:

- The MP brigade (I/R). The MP brigade (I/R) may augment the ASCC or the TSC during wartime. Its mission is to provide command, staff

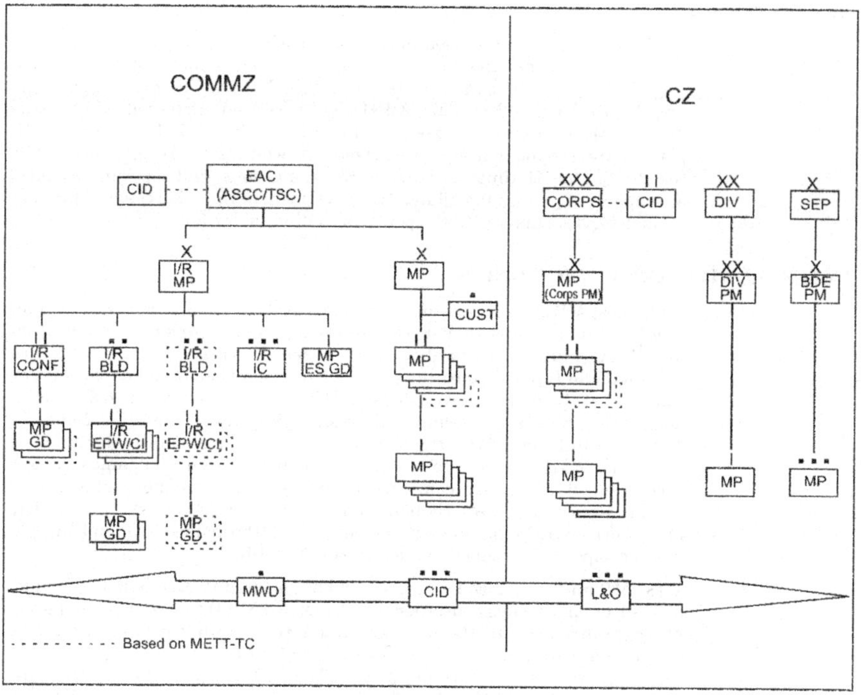

Figure 1-3. MP Structure in the TO

planning, and supervision of I/R operations. This includes coordination with joint and host-nation (HN) agencies, civilian police authority, nongovernmental organizations (NGOs), and US federal agencies.

- The MP brigade (CS). The MP brigade (CS) is assigned to the ASCC or the TSC during wartime (based on METT-TC). The MP brigade (CS) is capable of performing all five MP functions.
- The Criminal Investigation Division (CID) group. The CID group is a stovepipe organization that reports directly to the Commander, US Army Criminal Investigation Command (USACIDC). The CID group provides support to the ASCC and subordinate commands (TSC, corps, or division). See Chapter 9 for further discussion of CID support.

1-15. MP support to other EAC subordinate commands is performed only if MP resources are available. See Chapter 5 for further discussion of MP support to EAC.

SUPPORT IN THE CZ

1-16. MP support is provided in the CZ to each corps, division, and brigade (separate teams or initial/interim brigade combat teams [IBCTs]). An MP brigade (CS) is assigned to each corps, and the MP brigade commander is the corps's provost marshal (PM). A PM and his section, along with an organic division MP company, are assigned to each division. A PM cell and an MP platoon are organic to a separate brigade. A two-person PM cell is organic to the IBCT. The MP units assigned to corps, divisions, and separate brigades are capable of performing all five MP functions. They provide combat, CS, and combat-service-support (CSS) operations within their command's AO.

TYPES OF MILITARY POLICE UNITS

1-17. Most MP units supporting a TO and a JOA are capable of performing all five MP functions. However, the functions must be prioritized based on METT-TC and the availability of MP assets. Current MP structures are designed and tailored to better support the level of command deployed. For example, at the division level, division MP companies are organized as light, heavy, airborne, or air assault and are organic to their respective divisions. The EAC and corps MP brigades and battalions are equally designed to command and control a force mix of up to six battalions or companies. An MP escort-guard and guard company are designed to transport, guard, and provide security to EPWs, civilian internees (CIs), or dislocated civilians. The MP escort-guard company is assigned to the MP brigade (I/R), and the MP guard company is assigned to the MP battalion (I/R).

1-18. MP units can also be tailored and augmented to accomplish multiple, diverse, or specific missions. Customs, L&O, and MWD teams are examples of MP capabilities and flexible responses to a combatant commander's operational requirements. (See Table 1-1, pages 1-9 through 1-19, for a more complete description of MP units. See FM 19-10 and Army Regulation (AR) 190-12 for further information.) The battlefield workload analysis (BWA) is a tool used to determine the number of MP units required to perform some of these multiple missions (see Appendix C).

JOINT, MULTINATIONAL, AND INTERAGENCY OPERATIONS

1-19. In today's environment, the Army will rarely operate or fight alone. The high probability that the Army will operate in concert with its sister services, in an alliance with the forces of foreign nations, or in support of United Nations (UN) operations (when it is committed) is fully reflected in joint doctrine. In such operations, protecting LOC, key facilities, and command and control (C^2) centers will be a shared responsibility. Under this framework, MP units can expect to share the AO with joint, combined, multinational, or interagency resources. MP forces must be prepared to conduct a number of full-spectrum operations with a variety of government and nongovernmental agencies, other services, allied nations, and international agencies.

1-20. Corps and division commanders and staffs must plan (in advance) the transition from a single-service headquarters with joint representation to a joint headquarters capable of functioning as a joint task force (JTF) headquarters. When tasked to form a JTF headquarters, the corps or division must ensure that all of the staff sections and agencies have joint representation (see FMs 100-15 and 71-100). To this end, MP planners must ensure that the JTF is augmented with the appropriate MP forces and with the appropriate echeloned C^2.

1-21. Regardless of the force mix, the MP provide the force with unparalleled, multifunctional capabilities. Among these capabilities is the MP's ability to generate firepower or to handle populations such as EPWs/CIs, dislocated civilians, and refugees. Additionally, MP expertise in investigations and law enforcement enhances the capabilities of other joint, multinational, and interagency police and security forces.

1-22. MP security plans must reflect the joint synergy derived from combining the multiple and diverse capabilities of all participants. To capitalize upon that synergy, MP leaders must keep an open line of communication and coordination to offset the challenges presented by interoperability. Some of these interoperability challenges include—

- Differing political objectives.
- Differing capabilities.
- Cultural/language differences.
- Legal and policy constraints.
- Media impacts.
- Compromise of sensitive processes, procedures, and equipment.
- C^2.
- Communications (digital- and analog-equipment differences).

1-23. MP plans must also accommodate differences in planning capabilities, as well as differences in doctrine, training, and equipment. The intent is to match security missions with force capabilities. MP leaders must understand that operations will often involve multinational teams. While US forces routinely task-organize, this may be more difficult to accomplish with some multinational security forces. This kind of orchestration requires employing standardized procedures, communications, equipment, and liaison within the constraints of operations security (OPSEC).

1-24. Coordination is the key to mission accomplishment in multinational and interagency operations. A military coordination center or a civil-military operations center (CMOC) may meet this coordination requirement. The CMOC provides access for nonmilitary agencies desiring military (to include MP/CID) assistance and coordination. These nonmilitary agencies may include—

- Government organizations (GOs).
- NGOs.
- International organizations (IOs).
- International humanitarian organizations (IHOs).
- HN authorities and agencies.

1-25. The introduction of US Army MP forces in any joint, multinational, or interagency operation is based on METT-TC and the capabilities they bring to the operation. Effective integration of MP forces with other security forces reduces redundant functions, clarifies responsibilities, and conserves resources.

Table 1-1. Unit Descriptions

MP Unit	Mission	Action Elements	Support Role	Capabilities (Include But Are Not Limited to [See TOEs for Details])	Augmentation Needed for Mission Accomplishment
DIVISION MP UNITS					
MP Company (Airborne Division)	Perform the 5 MP functions.	A PM cell and 4 platoons. Each platoon has 3 squads and each squad has 2 three-man teams. Total: 24 three-man teams.	1 platoon deploys with each division brigade and provides DS during the assault phase. Then platoons revert to GS.	During the assault phase, 6 mounted, mobile or fixed-position teams deploy to provide DS for each brigade and 6 teams deploy to provide GS near the division main CP OR after the assault phase, 1 division EPW collection point (6 teams) and 18 mounted, mobile patrols or fixed-position teams provide GS OR 24 mounted, mobile or fixed-position teams are used in any combination.	The division band to provide close-in security for the division CP or to guard EPWs AND the corps CS MP to conduct combat operations within the division rear and provide sustained MMS, area security, special operations support, and escort for evacuating EPWs/CIs between division collection points and to the corps holding area AND the corps L&O MP and CID to conduct sustained L&O operations and criminal investigations AND division/corps transportation assets for evacuation of EPWs **NOTE: Both heavy division companies require corps CS MP augmentation for each forward support platoon (1 corps MP squad per forward support platoon).**
MP Company (Light Infantry Division)	Perform the 5 MP functions.	A PM cell and 3 platoons. Each platoon has 3 squads and each squad has 2 three-man teams. Total: 18 three-man teams	Provide GS for all units in the AO.	Provide screening security for the division main CP (6 teams), the division EPW collection point (6 teams), and 6 mounted, mobile or fixed-position teams.	
MP Company (Heavy Division)* *19333F000	Perform the 5 MP functions.	A PM cell and 2 division support platoons. Each platoon has 3 squads and each squad has a squad leader and 3 three-man teams AND 3 forward support platoons. Each platoon has 2 squads and each squad has a squad leader and 3 three-man teams. Total: 36 three-man teams	2 platoons provide GS and 3 platoons provide DS.	Provide screening security for the division main CP (6 teams), the division EPW collection point (6 teams), and 6 mounted, mobile or fixed-position teams AND EITHER 3 DS platoons providing their brigade with either 1 forward EPW collection point (3 teams) and 3 mounted, mobile or fixed-position teams or 6 teams in any combination OR 18 mounted, mobile or fixed-position teams are used in any combination.	
MP Company (Heavy Division)* *19333L000	Perform the 5 MP functions.	A PM cell and 5 platoons. Each platoon has 2 squads and each squad has 3 three-man teams. Total: 36 three-man teams	3 platoons provide GS and 3 platoons provide DS.	Provide screening security for the division main CP (6 teams), the division EPW collection point (6 teams), and 6 mounted, mobile or fixed-position teams AND EITHER 3 DS platoons providing their brigade with either 1 forward EPW collection point (3 teams) and 3 mounted, mobile or fixed-position teams or 6 teams in any combination OR 18 mounted, mobile or fixed-position teams are used in any combination.	

Table 1-1. Unit Descriptions (continued)

MP Unit	Mission	Action Elements	Support Role	Capabilities (Include But Are Not Limited to [See TOEs for Details])	Augmentation Needed for Mission Accomplishment
MP Company (Air Assault Division)	Perform the 5 MP functions.	A PM cell and 4 platoons. Each platoon has 3 squads and each squad has 2 three-man teams. Total: 24 three-man teams	Provide GS for all units in the AO.	24 mounted, mobile or fixed-position teams are used in any combination.	(See augmentation needs on previous page.)
MP COMBAT SUPPORT UNITS					
HHC MP Brigade	Provide C2 and coordinate the combat, CS, and CSS of all MP elements assigned or attached.	Command and staff elements and long-range planning section	Provide command, control, planning, and supervision for up to 6 MP battalions and other assigned or attached subordinate HQ.	Provide continuous command, control, coordination, planning, and supervision for subordinate units.	None
HHD MP Battalion (Combat Support)	Provide C2 and coordinate the combat, CS, and CSS of all MP elements assigned or attached.	Command and staff elements plus a support element	Provide command, control, planning, and supervision for up to 6 MP companies and other assigned or attached subordinate HQ.	Provide continuous command, control, coordination, planning, and supervision for subordinate units.	None
MP Company (Combat Support)	Perform the 5 MP functions.	4 platoons with 3 squads. Each squad has a squad leader and 3 three-man teams. Total: 36 three-man teams	Provide GS for all units in the AO.	Provide security for a unified or combined HQ or higher HQ or 1 main CP and 1 tactical CP OR operate 1 EPW holding area (9 teams) and 27 mounted, mobile or fixed-position teams for use in any combination.	Corps band assets to guard EPWs AND corps transportation for evacuation of EPWs AND external MP to escort EPWs
MP Company (Combat Support)	Perform the 5 MP functions.	4 platoons with 3 squads. Each squad has a squad leader and 3 three-man teams. Total: 36 three-man teams	Provide GS for all units in the AO.	Provide security for a unified or combined HQ or higher HQ or 1 main CP and 1 tactical CP OR operate 1 EPW holding area (9 teams) and 27 mounted, mobile or fixed-position teams for use in any combination.	Corps band assets to guard EPWs AND corps transportation for evacuation of EPWs AND external MP to escort EPWs
MP Company Arctic Support Augmentation Detachment	Augment MP units when operating in an arctic environment.	1 mechanic (63B10)	Provide mobility support to MP units.	Provide support as designated by the commander.	None
MP Detachment (C2)	Provide command and administrative personnel for MP law enforcement teams.	19503LA has a platoon HQS; 3 individuals. 19503LD has a commander and support personnel; 8 individuals.	Provide L&O. 19503LA provides platoon-level C2. 19503LD provides company-level command, control, and support.	Commands and controls law enforcement teams, investigations teams, MWD teams, and physical-security teams.	None

Table 1-1. Unit Descriptions (continued)

MP Unit	Mission	Action Elements	Support Role	Capabilities (Include But Are Not Limited to [See TOEs for Details])	Augmentation Needed for Mission Accomplishment
MP Detachment (Old Guard)	Perform the 5 MP functions.	4 squads. Each squad has 3 three-man teams. Total: 12 teams	Support the HHC old guard battalion.	Provide support as directed by the commander.	None
MP LAW AND ORDER (L&O) UNITS					
MP HQ Team	Provide C2, personnel administration, and logistical support to attached L&O augmentation teams.	Commander and support personnel	Provide L&O.	Provide C2, personnel administration, and logistical support to attached L&O augmentation teams.	None
MP Operations Team	Provide mission supervision, staff planning, and technical supervision required to support the L&O mission.	L&O operations officer and operations staff	Provide L&O.	Provide overall mission supervision, planning, employment, and coordination of support. Provide technical supervision and administrative support of MPI, TAI, and force protection MP. Provide an evidence custodian/police intelligence NCO to be responsible for confiscated property used in criminal offenses and to receive and collect the initial analysis of criminal/operational information/intelligence.	None
MP Desk or Desk and Record Team	Serve as the primary control point for L&O-related incidents and operations initiated by MP patrols.	2 three-man teams	Provide L&O.	Receive/record complaints, dispatch investigative persons, maintain control of offenders/detained persons, maintain the status of investigations, prepare reports, and provide criminal data to the police intelligence NCO for analysis.	None
MP Traffic Accident Investigations Team	Provide the technical capability to investigate traffic accidents.	1 two- or three-man team: 19517AD00 has 1 two-man team. 19523LE00 has 1 three-man team.	Provide L&O.	Augment MP capabilities to enforce MSR regulations; man checkpoints, roadblocks, and dismount points; patrol traffic areas; and perform traffic escort duties.	The 19517AD00 accident investigation team is dependent on the supported unit for transportation.
MP Investigation Team	Provide the technical capability to investigate criminal incidents and conduct surveillance operations.	One, two, or three-man teams: 19517AE has 1 two-man team. 19533LA has 1 one-man team. 19533LB has 1 one-man team. 19533LC has 1 three-man team.	Provide L&O.	Investigate criminal incidents, conduct surveillance, work with HN military and civilian police, and collect criminal and operational intelligence. 19533LA00 provides supervision for the teams. 19533LB00 provides investigative support for populations of not less than 700 troops. 19533LC00 provides investigative support for populations of not less than 2,100 troops.	Dependent on the supported unit for transportation

Table 1-1. Unit Descriptions (continued)

MP Unit	Mission	Action Elements	Support Role	Capabilities (Include But Are Not Limited to [See TOEs for Details])	Augmentation Needed for Mission Accomplishment
MP Force Protection and Physical-Security Teams	Provide the expertise and technical capability to assist units in safeguarding personnel, equipment, and facilities.	One, two, or three-man teams: 19517AF has 1 two-man team. 19593LA has 1 three-man team. 19593LB has 1 one-man team.	Provide L&O and area security.	Conduct vulnerability assessments and identify and prioritize critical facilities and key terrain within the AO. Identify mission-essential activities that are vulnerable to criminal acts or disruptive activities. Ensure that activities are inspected to determine if safeguards are adequate. Assist base/base cluster commanders with the development of internal defense plans.	The 19517AF is dependent on the 19517AB operations team for transportation.
MP Detach-ment (Patrol Supervision Team)	Provide supervision for 3 MP teams.	1 MP NCO	Provide L&O.	Plan, direct, and supervise the employment of assigned or attached MP teams.	Dependent on the supported unit for transportation
MP Detach-ment (Motor and Dismounted Patrol Team)	Perform L&O operations.	1 three-man team	Provide L&O.	Protect designated personnel or facilities. Perform route and area reconnaissance, enforce MSR regulations, provide refugee and straggler control, and disseminate information. Perform peacekeeping operations to maintain surveillance over an area, observe activities, and report findings. Preserve or establish L&O. Apprehend absentees or deserters (US military personnel) in conjunction with civil law enforcement agencies.	The dismounted patrol team is dependent on the supported unit for transportation, when required.
MP Detach-ment (Registration Team)	Perform vehicle and firearm registration.	1 three-man team	Provide L&O.	Service up to 10,000 personnel for registration of individuals, vehicles, and firearms on a closed post or area. Issue credentials prescribing limits of circulation and privileges. Process up to 50 personnel daily for fingerprints and photography services.	Dependent on the supported unit for transportation
Senior Military Customs Inspector Supervisor	Perform technical supervision, staff planning, and coordination for customs inspection teams.	1 MP NCO	Provide L&O.	Supervise two senior military customs teams to ensure that personnel, equipment, and material meet customs, immigration, Department of Agricul-ture, and other federal agency requirements for units and personnel redeploying to the US.	None

Table 1-1. Unit Descriptions (continued)

MP Unit	Mission	Action Elements	Support Role	Capabilities (Include But Are Not Limited to [See TOEs for Details])	Augmentation Needed for Mission Accomplishment
Senior Military Customs Inspection Team	Conduct customs inspections and train and supervise redeploying unit personnel to augment US customs inspectors.	1 five-man team	Provide L&O.	Conduct inspections and advise redeploying units. Train selected individuals from the redeploying unit to augment US customs efforts. NOTE: One team supports one port of embarkation.	The redeploying unit undergoing customs inspection, to augment US Customs efforts. The number of personnel is proportional to the size of the unit.
MILITARY WORKING DOG (MWD) UNITS					
MWD Kennel Master Team	Supervise MWD teams.	One- or two-man team. 19537AA has 1 two-man team. 19583LG has 1 one-man team.	Provide L&O, area security, and I/R operations support.	Supervise, plan, and coordinate MWD operations and support requirements. Provide MWD handler/canine proficiency certification.	None
MWD Explosive, Narcotics, and Patrol Team	Detect explosives and controlled substances and search for, detect, and control personnel in support of crime scene searches. Provide personal protection, MOUT, health and welfare inspections, and I/R and customs operations.	One-man/one-dog team. The 19537A000-series TOEs have 3 teams each. The 19583L000-series TOEs have 1 team each.	Provide L&O, area security, and I/R support.	The 19537A-series dog teams are capable of providing a 24-hour explosive, narcotics, and tracking handler/dog requirement or three concurrent short-duration missions each requiring 1 handler/dog team. The duration of the missions will vary based on the climate, the environment, and the individual dog's ability. The 19583L-series dog teams provide short- duration missions based on the dog's ability. Explosive and narcotics teams provide MWD patrol support when not employed with explosive and narcotics detection.	None
MILITARY POLICE INTERNMENT/RESETTLEMENT (I/R) UNITS					
MP Command (I/R)	Provide command, control, staff planning, and supervision of I/R operations performed by all assigned or attached elements.	Command and staff elements	Provide I/R operations support.	Provide command, control, and staff planning for 2 or more MP brigades.	MP IRIC AND MP EPW evacuation detachment

Table 1-1. Unit Descriptions (continued)

MP Unit	Mission	Action Elements	Support Role	Capabilities (Include But Are Not Limited to [See TOEs for Details])	Augmentation Needed for Mission Accomplishment
MP Brigade (I/R)	Provide command, control, and staff planning for I/R operations performed by all assigned or attached MP elements.	Command and staff elements	Provide I/R operations support.	Provide command, control, and staff planning for 2 to 7 MP I/R battalions or up to 21 MP I/R battalions when augmented by 7 brigade liaison detachments (BLDs). Plan and provide staff supervision of I/R collection and evacuation operations. Coordinate with HN military territorial organizations and civilian police authorities, nongovernment organizations, private volunteer organizations, and US federal agencies on I/R matters. Provide coordination and support for out-of-theater evacuation of EPWs, if required. Support posthostilities operations.	MP IRIC AND MP EPW evacuation detachment AND MP I/R BLD (TOE 19453LD00) **NOTE: The BLD expands the staff planning and coordination capabilities on a ratio of 1 BLD to 3 MP I/R battalions.**
MP Battalion (I/R)	Provide command, staff planning, administration, and logistical support for the operation of an I/R facility.	Command and staff elements	Provide I/R operations support.	Operate an internment facility for either EPWs/CIs or DCs or a confinement facility for US prisoners. (Never more that one category at the same time.)	MP detachment (I/R) (EPW/CI) for EPW/CI/DC internment missions AND MP detachment (I/R) (confinement) for US prisoner confinement missions AND MP company (EG) for security of the evacuation and/or movement of US prisoners or EPWs/CIs/DCs AND MP company (guard) for prisoner guard services AND Corps/EAC transportation assets for movement of EPWs/CIs/DCs

Table 1-1. Unit Descriptions (continued)

MP Unit	Mission	Action Elements	Support Role	Capabilities (Include But Are Not Limited to [See TOEs for Details])	Augmentation Needed for Mission Accomplishment
MP Detach-ment (I/R) (EPW/CI)	Augment the MP battalion (I/R) to provide supervision, administration, combat health support, and logistical support for operating an internment facility interning EPWs/CIs or housing DCs.	2 compound control sections (2 three-man teams), a work project section (1 two-man team), a personnel section, and a supply section.	Provide I/R operations support.	Provide command, control, and supervision of up to 1,000 EPWs/CIs or 2,000 DCs. Provide supply and subsistence support (to include supervision of food preparation) for up to 1,000 EPWs/CIs or 2,000 DCs. Direct the activities related to assigning and supervising work projects for EPWs/CIs. Augment the MP I/R battalion in processing and maintaining records for up to 1,000 EPWs/CIs or 2,000 DCs.	MP company (guard) for prisoner guard services
MP Company (Escort Guard)	Provide supervisory and security personnel for evacuating and/ or moving EPWs/CIs.	4 platoons with 3 squads each. Each squad has a squad leader and 3 three-man teams. Total: 36 teams	Provide I/R operations support.	Provide security for the movement of the following numbers of POWs or CIs by the methods indicated: Marching - 1,000 to 1,500. Vehicle - 1,500 to 2,000. Rail - 2,000 to 3,000. Air - 2,125 to 2,406.	Corps/EAC transportation assets for movement of EPWs/CIs
MP Company (Guard)	Provide guards for EPWs/CIs or US prisoners, installations, and facilities.	3 platoons with 3 squads each. Each squad has a squad leader and 2 five-man teams Total: 18 teams.	Provide I/R operations support.	Provide security for a confinement facility containing up to 500 US military prisoners. Provide guards for securing 1 EPW/CI compound containing up to 2,000 EPWs/CIs or 4,000 DCs. Provide security guards for 3 railway terminals, each having up to 8 tracks. (Includes EPW/CI rail movement operations and protection of sensitive material within the terminal.) Provide guards for 1 military installation or facility up to 240,000 square yards in size, containing sensitive material. (This unit can provide guards for material transit.)	None
MP I/R Processing Squad	Process EPWs/ CIs.	1 squad leader and 1 eight-man processing squad	Provide I/R operations support. Augment the I/R battalion processing capability to meet operational requirements.	Provide processing capabilities of about 8 EPWs/CIs per hour.	None

Table 1-1. Unit Descriptions (continued)

MP Unit	Mission	Action Elements	Support Role	Capabilities (Include But Are Not Limited to [See TOEs for Details])	Augmentation Needed for Mission Accomplishment
I/R Brigade Liaison	Expand the MP brigade (I/R) or ASCC C2 capabilities.	Liaison officer and support staff	Provide I/R operations support.	When assigned to the MP I/R brigade, provide staff augmentation that expands the brigade's staff planning, coordination, and C2 capabilities for 3 MP I/R battalions. When assigned to the ASCC, provide I/R staff augmentation and a liaison link with allied/HN forces to ensure that the care and handling of US captured EPWs/CIs is in compliance with the Geneva Conventions.	None
I/R Camp Liaison Team	Provide continuous accountability of EPWs/CIs captured by US forces and transferred to an allied/HN for internment.	Liaison officer and support staff	Provide I/R operations support.	Provide advice, as requested, to commanders and staffs of allied/HN-operated internment facilities. Verify arrival, forward records, and provide continuous accountability for US captured EPWs/CIs interned in allied/HN facilities. Monitor prisoner treatment to ensure compliance with the Geneva Conventions. Receive/certify allied/HN requests for reimbursement of expenses associated with interning EPWs/CIs captured by US forces.	None
I/R Processing Liaison Team	Process and verify the transfer of EPWs/CIs captured by US forces to a allied/HN facility.	Liaison officer and support staff	Provide I/R operations support.	Provide processing and transfer capability for US-captured prisoners to an allied/HN EPW/CI facility.	None
MP Detachment (EPW Evacuation)	Provide evacuation support of EPWs.	EPW coordination officer and staff plus 10 one-man POD/POE teams and 3 two-man POE teams	Provide I/R operations support.	Coordinate administrative, logistical, and transportation support for up to 7 PODs and 6 POEs used for out-of-theater EPW evacuation and the escort guard personnel that support the movements. Coordinate for theater security at PODs/POEs and intermediate refuel points.	Responsible POD/POE security forces

Table 1-1. Unit Descriptions (continued)

MP Unit	Mission	Action Elements	Support Role	Capabilities (Include But Are Not Limited to [See TOEs for Details])	Augmentation Needed for Mission Accomplishment
MP Detachment (Internment Resettlement Information Cener [IRIC])	Provide a US central agency in each theater of war to receive, process, maintain, disseminate, and transmit the required information and data relating to EPWs, American POWs, CIs, and DCs within the theater.	Command and staff personnel	Provide I/R operations support. Serve as the single source for collection and storage of EPW/CI information in theater. Forward information to the National Prisoner of War Information Center at DA.	Collect, process, and disseminate (to authorized agencies) information regarding interned personnel detained in theater and those released to the custody of allied/HN authorities within theater. Receive, document, and disseminate to the theater commander information received regarding American POWs, CIs, and foreign nationals who are captured, missing, or otherwise detained. Receive, store, and dispose of personal property belonging to interned personnel who have died, escaped, or been repatriated and any property belonging to enemy soldiers killed in action that is not disposed of through grave registration channels.	None
MP Detachment (I/R) (Confine-ment)	Augment an MP battalion (I/R) in operating a confinement facility. Provide supervision, administration, combat health support, and logistical support of US military prisoners.	Corrections officer and confine-ment facility staff	Provide I/R operations support.	Provides a control team for the command, control, and supervision of US military prisoners and a disci-plinary guard team to supervise custodial personnel for a maximum custody of 500 prisoners. Provide administration, health service, supply, and food ser-vice personnel to augment the battalion in providing these services for internees. (The food service per-sonnel supervise and train US military prisoners working in the internee dining facility.)	MP battalion (I/R) for command, staff, plan-ning, and operational support AND MP company (guard) for prisoner guard services

Table 1-1. Unit Descriptions (continued)

MP Unit	Mission	Action Elements	Support Role	Capabilities (Include But Are Not Limited to [See TOEs for Details])	Augmentation Needed for Mission Accomplishment
MP Confinement Facility Detachment	Provide C2, staff planning, administration, and logistical support for a confinement facility for US military prisoners.	Commander and facility staff	Provide I/R operations support.	Provide command, control, and administrative support for a confinement facility.	MP guard company to provide exterior perimeter security and required guard support to the confinement facility AND MP correctional team supervisor (TOE19553LE) to provide supervision of the guard force AND MP correctional team work supervisor (TOE 19553LF) to provide supervision of prisoners
MP Correctional Team Supervisor	Provide correctional supervision of prisoner work guard forces within a confinement facility.	2 two-man teams	Provide I/R operations support.	Provide correctional supervision to guard force personnel and US military prisoners within a modular-configured confinement facility.	None
MP Correctional Team Work Supervisor	Provide direct control and supervision of military prisoners within a confinement facility.	1 two-man team	Provide I/R operation support.	Provide correctional control, supervisory functions, and escort duties for military prisoners within a modular-configured confinement facility.	None
MILITARY POLICE CRIMINAL INVESTIGATION (CID) UNITS					
MP Group (CID)	Provide command, control, staff planning, and supervision for all CID elements within the theater.	Command and staff elements	Provide L&O.	Provide command, control, staff planning, and supervision over all CID elements within the theater. **NOTE: The Criminal Investigation Command is a stovepipe organization.**	None

Table 1-1. Unit Descriptions (continued)

MP Unit	Mission	Action Elements	Support Role	Capabilities (Include But Are Not Limited to [See TOEs for Details])	Augmentation Needed for Mission Accomplishment
MP Battalion (CID)	Provide command, control, staff planning, and supervision for all assigned or attached CID elements.	Command and staff elements	Provide L&O.	Provides command, control, staff planning, and supervision over all assigned and attached CID elements.	None
MP Detachment (CID) HQ Cell	Provide C2, evidence custody control, and investigative administration support. Coordinate for personnel administration and logistic support.	Commander and support staff	Provide L&O.	Provide C2 of assigned CID SA sections and/or teams performing all CID operations in their area of responsibility.	None
MP Detach-ment (CID)	Provide criminal investigative support to Army commanders at all echelons.	2-man teams. The number of teams varies by organization and is as follows: DSE heavy - 4 DSE light - 3 Section A - 4 Section B - 4 Supervisory team - 1 Senior team - 1 SA team - 1	Provide L&O.	Supervise and conduct criminal investigations.	When directed, MP CS, L&O, and MWD support

Chapter 2
Battle Command

Command is the authority a commander in military service lawfully exercises over subordinates by virtue of rank and assignment. Leaders possessing command authority strive to use it with firmness, care, and skill.

FM 101-5-1

Battle command is the exercise of command in an operation against a hostile, thinking opponent. Battle command includes visualizing the current state and the desired end state, then formulating concepts of operations to get from one state to the other at the least cost. In addition to visualizing and formulating concepts, battle command encompasses assigning missions; prioritizing and allocating resources; selecting the critical time and place to act; and knowing how and when to make adjustments in the fight. Battle command enables MP commanders to lead, prioritize, and allocate assets required in support of the Army commander. MP commanders must observe, orient, decide, and act on their decisions quickly. Information is the key element in the battle-command process; therefore, the commander must have accurate and timely information upon which to base his decisions.

OVERVIEW

2-1. The battle command of MP units is typically decentralized due to the nature of their CS functions, METT-TC, and the needs of the Army commander. This places the burden of sound, timely decision making to the lowest levels. MP leaders must develop a keen sense of situational awareness and visualization, and they must constantly track the actions of supported units.

BATTLEFIELD VISUALIZATION

2-2. The ability to visualize the battlefield is a critical element of battle command. Battlefield visualization is an essential leadership attribute and is critical to accomplishing the mission. It is learned and attained through training, practice, experience, technical and tactical knowledge, and available battle-command technologies. It results when the MP commander understands the higher commander's intent, his assigned mission, the enemy, and the friendly force's capabilities and limitations. See Appendix D for further information on command technologies.

2-3. Battlefield visualization includes the MP commander's view of what his forces will do and the resources needed to do the mission. He envisions a sequence of actions that will cause his MP forces to perform at the desired end state. Ultimately, the MP commander's battlefield vision evolves into his intent and helps him develop his concept of operations.

COMMANDER'S INTENT

2-4. The commander's intent is a key part of Army orders. It is a clear, concise statement of what the force must do to succeed with respect to the enemy, the terrain, and the desired end state. It provides the link between the mission and the concept of operations by stating key tasks. These tasks, along with the mission, are the basis for subordinates to exercise initiative when unanticipated opportunities arise or when the original concept of operations no longer applies. MP leaders at all echelons must ensure that the mission and the commander's intent are understood two echelons down (see FM 101-5).

CONCEPT OF OPERATIONS

2-5. The commander's intent does not include the method by which the MP units will accomplish the mission. This method is called the commander's concept of operations. It must—

- Convey the commander's vision of how to accomplish the mission in a manner that allows his subordinates maximum initiative.
- Build around intelligence gathering and the precise employment of MP resources.
- Provide the basis for task organization, scheme of maneuver, terrain organization, tasks to subordinates, and synchronization.

COMMAND AND CONTROL RELATIONSHIPS

2-6. MP units are assigned to, attached to, or placed under the operational control (OPCON) of MP or other units they support. OPCON is the authority to perform command functions over subordinate forces. This includes organizing and employing commands and forces, assigning tasks, designing objectives, and giving authoritative direction necessary to accomplish the mission. MP C^2 relationships may be changed briefly to provide better support for a specific operation or to meet the needs of the supported commander. MP units may be placed under the OPCON of another unit commander for short-term operations. The MP unit remains in this relationship only as long as it is needed for that operation.

> MP support to the Bosnian municipal elections consisted of one division and two corps MP companies. These MP assets, attached to Task Force (TF) Eagle, were task-organized from different sources. The division MP company and the PM cell were organic to TF Eagle's mechanized infantry division headquarters, but the two corps MP units were from US Army Forces Command (FORSCOM) MP battalions in CONUS.

SUPPORT RELATIONSHIPS

2-7. MP units on the battlefield provide two types of support—general support (GS) and direct support (DS). Corps and EAC MP units provide GS to their respective corps/EAC subordinate commands. Light, airborne, and air-assault MP companies provide GS to their respective divisions. Heavy-division MP companies provide GS to the division rear and DS to the division's subordinate brigades.

STAFF RELATIONSHIPS

2-8. The PM for each level of command is that command's advisor on MP combat, CS, and CSS operations. The PM—

- Advises the commander and staff about MP abilities/capabilities.
- Supervises the preparation of plans and dictates policies.
- Coordinates MP operations.
- Assists and supervises the interaction of supporting and supported units.
- Reviews current MP operations.
- Coordinates with allied forces and HN military and civil police.
- Ensures that MP plans and operations supporting the commander's tactical plan are carried out.
- Recommends when and where to concentrate the command's MP assets.
- Supervises or monitors MP support in the command's AO.

2-9. The PM works daily with the commander and staff officers who employ MP resources and whose AORs influence MP support. The PM works closely with the coordinating staff at the appropriate command level to coordinate MP support. He ensures that MP planning is practical and flexible, that plans are coordinated with staff sections and subordinate commands, and that plans reflect manpower and resources needed by MP. (This includes the need for C^2, fire support, equipment, and supplies. It also includes construction, communication, transportation, and aviation support.) As new information is received, the PM reviews, updates, and modifies the plans. He ensures that the echelon commander gets the necessary MP support.

2-10. In the absence of specific directions or orders, the PM plans the use of MP assets. He evaluates the current operations and projects the future courses of action (COAs). He bases his plans on assumptions consistent with the commander's intent and a thorough knowledge of the situation and mission. The PM considers—

- METT-TC.
- Current estimates developed by the intelligence preparation of the battlefield (IPB) and the police information assessment process (PIAP).
- The environment within the AO. This includes the climate, the terrain, and obstacles. It also includes the legal authority and status of the force; the width, depth, size, and location of built-up areas; and the attitudes and abilities of the local populace.

- The types of units operating in the area (to include joint, combined, multinational, and interagency units) and the missions and capabilities of these units. This knowledge is imperative to understand their capability to counter threats in their area.
- The specific missions of MP units in the area and the impact that rear-area security operations will have on the ability of these units to perform other functions.
- Personnel, vehicles, and equipment in the MP units.

2-11. Coordination and communication between the PM and Army commanders is essential. Such actions ensure timely and efficient MP support to all levels of command during any operation. The informal, technical chain of coordination is an open line of communication between PMs at different echelons. The informal chain of coordination fosters cooperation and help among the MP elements at each echelon. For instance, when the division PM needs more assets to accomplish added missions, he initiates coordination with the corps PM. If the corps PM can provide support, the division PM formalizes his request for assistance through the division Assistant Chief of Staff, G3 (Operations and Plans) (G3).

Chapter 3

The Threat

In the 40-odd years of the Cold War, in many locations around the world, the Army performed a deterrent role as part of the containment strategy. In other places, at other times, the Army fulfilled the Nation's expectation in operations too small to be called "wars," although no less dangerous. To the soldier on the ground, Operations Urgent Fury in Grenada and Just Cause in Panama were indistinguishable from combat operations of their forefathers. Operations Provide Comfort in Iraq and Restore Hope in Somalia, although peace operations, also proved to be dangerous.

FM 100-1

The end of the Cold War has reduced, but not eliminated, the most immediate threat to the security of the US and other western nations. However, the absence of a dominant, identifiable threat has produced a far more complex and confusing strategic environment than the one that was present during the Cold War. Forward-deployed and CONUS-based ARFOR and civilians are and will continue to be engaged in a range of military actions. These actions stem from deterring conflicts to conducting peacetime engagement operations to providing support to civil agencies at home and abroad.

OVERVIEW

3-1. During the past decade, the US has deployed forces in multiple operations that have included crisis response in combat situations as well as participation in noncombat activities. The Army's presence in South Korea, Saudi Arabia, and Kuwait and its deployments to Somalia, Haiti, Bosnia, and Kosovo are clear indicators that the military must be prepared to face not only the traditional threat, but also a nontraditional, nonecheloned enemy. To support Army commanders successfully, MP leaders must understand the nature and complexity of these threats and how they can potentially affect the desired strategic, operational, and tactical end states.

REAR-AREA AND SUSTAINMENT OPERATIONS

3-2. The rear area for any particular command is the area extending forward from its rear boundary to the rear of the area assigned to the next lower level of command. This area is provided primarily for the performance of support functions. Operations in the rear area assure freedom of action and continuity of operations, sustainment, and C^2. Sustainment operations are those that enable shaping and decisive operations by assuring freedom of action and continuity of operations, CSS, and C^2 (see FM 3-0). Sustainment operations include the following elements:

- CSS.
- Rear-area and base security.
- Movement control.
- Terrain management.
- Infrastructure development.

3-3. During the Cold War, the danger to rear areas included forces that would be deployed in support of major soviet-style operations. The adversaries using the soviet model could be expected to engage in intense combat activity in their enemy's rear area. Their forces were prepared to penetrate into the enemy's rear and to attack and destroy its reserve forces and rear-area installations. To protect the rear areas, the MP were among the first mobile fighting forces available to the battlefield commander and thus, a source of combat power. Today, the Army commander uses the MP's flexibility and their modular-force training, adaptability, and mobility to serve as a combat multiplier throughout his entire AO. During sustainment operations, the MP perform all functions to ensure freedom of maneuver in support of the overall operational effort.

3-4. Failure to protect our forces during sustainment operations normally results in failure of the entire operation. Sustainment operations determine how fast ARFOR reconstitute and how far they can exploit success. The likelihood of MP units encountering the enemy and engaging in direct combat (not only in the rear area, but also during sustainment operations) cannot be underestimated.

3-5. Threats to rear-area and sustainment operations exist throughout the full spectrum of military operations. These threats may be related or independently engaged, but their effects are frequently cumulative. Threats to rear-area and sustainment operations are usually theater-dependent and are not limited to those outlined in this manual. Joint Publication (JP) 3-10 further discusses the threat in the rear area. Although JP 3-10 defines the threat in the context of a JRA, MP leaders can expect the same level of activity anywhere that US forces are deployed.

RECEPTION, STAGING, ONWARD MOVEMENT, AND INTEGRATION OPERATIONS

3-6. Reception, staging, onward movement, and integration (RSOI) operations consist of essential and interrelated processes in the AO that transform arriving personnel and materiel into forces capable of meeting operational requirements. During RSOI operations, the threat encountered will depend mostly on the type of entry, the nature of the operation, and the enemy. During major contingencies, forces deploy from power-projection platforms within the US or forward bases. The PM must plan MP support during the initial stages of the deployment to ensure the protection of follow-on forces and the detection of potential threats (see FM 100-17-3).

3-7. MP support to RSOI operations includes, but is not limited to—

- Conducting AS operations to counter or prevent enemy actions against marshalling and staging areas.
- Conducting convoy, airport, and rail security operations.
- Conducting populace- and resource-control operations.
- Conducting other physical-security and force-protection measures.
- Conducting other MP functions (as determined by the PM).

THREAT LEVELS

3-8. The threat is divided into three levels. These levels provide a general description and categorization of threat activities, identify the defense requirements to counter them, and establish a common reference for planning guidelines. MP leaders must understand that this does not imply that threat activities will occur in a specific sequence or that there is a necessary interrelationship between each level.

LEVEL I

3-9. Level I threats include the following types of individuals or activities:

- **Enemy-controlled agents.** Enemy-controlled agents are a potential threat throughout the rear area. Their primary missions include espionage, sabotage, subversion, and criminal activities. Their activities span the range of military operations and may increase during both war and military operations other than war (MOOTW). These activities may include assassinating or kidnapping key military or civilian personnel or guiding special-purpose individuals or teams to targets in the rear area.
- **Enemy sympathizers.** Civilians sympathetic to the enemy may become significant threats to US and multinational operations. They may be the most difficult to neutralize because they are normally not part of an established enemy-agent network, and their actions will be random and unpredictable. During war and MOOTW, indigenous groups sympathetic to the enemy or those simply opposed to the US can be expected to provide assistance, information, and shelter to guerrilla and enemy unconventional or special-purpose forces operating in the rear area.
- **Terrorism.** Terrorists are among the most difficult threats to neutralize and destroy. Their actions span the full spectrum of military operations.
- **Civil disturbances.** Civil disturbances, such as demonstrations and riots, may pose a direct or indirect threat to military operations. Although this threat may not be of great impact during war, it may significantly change and affect MOOTW.

LEVEL II

3-10. Level II threats include the following types of forces:

- **Guerilla forces.** Irregular and predominantly indigenous forces conducting guerrilla warfare can pose a serious threat to military forces and civilians. They can cause significant disruptions to the orderly conduct of the local government and services.

- **Unconventional forces.** Special-operations forces (SOF) are highly trained in unconventional-warfare techniques. They are normally inserted surreptitiously into the rear area before the onset of an armed conflict. They establish and activate espionage networks, collect intelligence, carry out specific sabotage missions, develop target lists, and conduct damage assessments of targets struck.
- **Small tactical units.** Specially organized reconnaissance elements are capable of conducting raids and ambushes in addition to their primary reconnaissance and intelligence-gathering missions. Small (size or capability), bypassed conventional units, as well as other potential threat forces, are also capable of conducting raids and ambushes to disrupt operations.

LEVEL III

3-11. Level III threats are made up of conventional forces. Potential threat forces are capable of projecting combat power rapidly by land, air, or sea deep into the rear area. Specific examples include airborne, heliborne, and amphibious operations; large, combined-arms, ground-forces operations; and bypassed units and infiltration operations involving large numbers of individuals or small groups infiltrated into the rear area, regrouped at predetermined times and locations, and committed against priority targets. Level III forces may use a combination of the following tactics as a precursor to a full-scale offensive operation:

- **Air or missile attack.** Threat forces may be capable of launching an air or missile attack throughout the rear area. It is often difficult to distinguish quickly between a limited or full-scale attack before impact; therefore, protective measures will normally be based on the maximum threat capability.
- **Nuclear, biological, and chemical (NBC) attack.** Commanders must be aware that NBC munitions may be used in conjunction with air, missile, or other conventional-force attacks. The NBC weapons could also be used at Level I or II by terrorists or unconventional forces in order to accomplish their political or military objectives.

THREAT-LEVEL MATRIX

3-12. Table 3-1 lists the threat levels and their likely appropriate responses. The threat levels listed are based on the type of threat. The table should not be construed as restricting the response options to any particular threat.

THREAT PRIORITIES

3-13. The threat will attempt to perform the following operations against targets in the rear area:

- Detect and identify targets.
- Destroy or neutralize operational weapons-system capabilities.
- Delay or disrupt the timely movement of forces and supplies.

Table 3-1. Threat Levels

Threat Level	Example	Response
I	Agents, saboteurs, sympathizers, and terrorists	Unit, base, and base-cluster self-defense measures
II	Small tactical units, unconventional-warfare forces, guerrillas, and bypassed enemy forces	Self-defense measures and response forces with supporting fires
III	Large tactical-force operations (including airborne, heliborne, amphibious, infiltration, and bypassed enemy forces)	Timely commitment of a TCF

- Weaken the friendly force's C^2 network.
- Disrupt support to combat forces.
- Set the stage for future enemy operations.
- Create panic and confusion throughout the rear area.

3-14. Typical examples of enemy priority targets include the following:

- NBC-weapons storage sites and delivery systems.
- Key command, control, communications, computers, and intelligence (C4I) facilities.
- Air-defense artillery (ADA) sites.
- Airfields and air bases.
- Port facilities.
- Main supply routes (MSRs) and MSR checkpoints.
- Key LOC.
- Reserve assembly areas (AAs).
- Troop barracks.
- Critical civilian and logistics facilities.

THREAT LOCATION

3-15. The fact that the Cold War has ended does not imply that our traditional threat has ended. North Korea and Iraq are constant reminders of this fact. For the near future, Army commanders will fight units with Cold-War-era equipment and tactics. The Army trains and is prepared to fight an enemy capable of interfering with our freedom of maneuver throughout the battlefield. On an extended battlefield with asymmetric threats, the danger to high-value assets (HVAs) (including CSS, C^2, communication nodes, and MSRs) only increases. The idea that the danger to the rear area decreases as you travel farther away from the forward edge of the battle area (FEBA) is not true. Threat intensity does not depend on geographical location; it depends on what operations the enemy believes must be initiated (and to what degree) to achieve its objective in the rear area. Military commanders depend on the MP to delay and defeat threats in their AO with a mobile reaction force.

COMMUNICATIONS ZONE

3-16. The nature of the COMMZ will encourage Level I and II threats to concentrate along the LOC and other areas of military significance. MP units will encounter an enemy that is capable of disrupting operations throughout the COMMZ while employing terrorist activities, enemy-controlled agent activities, enemy sympathizers, and saboteurs. If the enemy is Level III capable, MP leaders must expect infiltrations and air, missile, or NBC attacks as a precursor to a major Level III operation.

CORPS REAR AREA

3-17. The activities in Levels I and II will be similar in composition and density as in the COMMZ, but they will target key corps units, key facilities, and corps sustainment capabilities. The threat activities, especially at smaller unit levels, may even precede hostilities. MP leaders must be alert and prepared to encounter unconventional forces conducting diversionary or sabotage operations and small combat units conducting raids, ambushes, or reconnaissance operations or collecting special warfare intelligence. With the fast tempo of offensive operations, MP leaders must also be alert and prepared to encounter bypassed forces that can disrupt operations in the corps rear area.

DIVISION REAR AREA

3-18. The division rear area (DRA) contains many types of CS and CSS units and conducts many complex operations. As in the COMMZ and the corps rear area, the full spectrum of Level I, II, and III activities may occur in the DRA. The main target will be the division's HVA (including key C^2 facilities; airfields; artillery, aviation, and air-defense assets; LOC; and essential CSS units). The threat may conduct diversionary attacks, sabotages, raids, ambushes, and reconnaissance operations to affect the commander's freedom of maneuver and the continuity of operations. Unlike corps MP, the likelihood of division MP encountering bypassed enemy forces is expected. Failure to delay or defeat these forces will impact division operations.

OTHER TYPES OF THREATS

3-19. As US forces are deployed throughout the world, they will have to face nontraditional, asymmetric threats (other than those listed in Table 3-1, page 3-5) that may be geographically specific. As part of situational awareness, and in coordination with military intelligence (MI) and CID personnel, MP leaders must evaluate and assess the impact of these threats in their AO. A TO is vulnerable to any or a combination of the following threats:

- National or international organized crime.
- Narcotics traffickers.
- Narcotics terrorists.
- Extremist groups.
- Paramilitary groups.
- Ethnic or religious disputes.
- Trade in illegal weapons or strategic materials.

3-20. MP leaders must be aware that other threats exist and that they have the same potential as the Level I and II threats to disrupt operations in rear-area or sustainment operations. In some instances, the above threats' capabilities or the massing of personnel may have the same potential threat as a Level III threat.

COUNTERING THE THREAT

3-21. The disruption of rear-area and sustainment operations directly affects military efforts. Three types of forces may be used to counter the threat in these areas—a base/base-cluster self-defense force, a response force, or a tactical combat force (TCF).

BASE/BASE-CLUSTER SELF-DEFENSE FORCE

3-22. A base cluster is established when the appropriate echelon rear-operations cell or command post (CP) places geographically contiguous or noncontiguous bases under the control of a headquarters. The base cluster becomes the next higher tactical C^2 headquarters of those bases. The rear-operations cell or the rear CP may also establish a base cluster for a corps support group (CSG), an area support group (ASG), or other CSS units operating in the corps or division rear areas.

3-23. US ARFOR have the inherent responsibility to contribute as many forces as possible for base defense and local security for themselves and their facilities, installations, and activities. Each base and base-cluster commander must develop a defense plan to detect, defeat, and minimize the effects of Level I and limited Level II threat attacks on his base or base cluster (including NBC attacks). To maximize the unit's mission accomplishment, defense plans must be flexible and allow for differing degrees of security based on the probability of threat activity. Defense plans are given to MP units operating near the base or base cluster. The base commander most often employs a series of defense measures providing internal and perimeter security. His internal reaction forces use organic weapons to neutralize and defeat most low-level threat activity. Although not fully equipped to engage major conventional or unconventional enemy forces that may confront him, a base commander must deploy his personnel to defend themselves until MP, HN, local police, or combat forces (if available) can respond.

RESPONSE FORCE

3-24. A response force is summoned when the base or base cluster is faced with threat forces that are beyond their self-defense capability. If the MP are the designated response force, they must—

- Coordinate with the supported bases or base-cluster commanders to conduct a joint IPB.
- Review base and base-cluster self-defense plans.
- Exchange signal-operating-instructions (SOI) information.
- Identify MP contingency plans to counter likely enemy activities.
- Integrate ADA, engineer, chemical, field-artillery (FA), Army-aviation, and close-air-support (CAS) fire support into their plans (if available).

3-25. MP units help the base or base cluster return to its primary mission by defeating Level II threats. MP units closely watch likely avenues of enemy approach, possible landing zones (LZs), drop zones (DZs), C^2 facilities, and other key installations. They accomplish this through the MMS, AS, and PIO functions.

3-26. If MP units are not the designated response force, they may become the initial response force for units within their AO. When this occurs, they block, delay or, if possible, destroy enemy elements within their capability. If the attack is by a larger or more capable force, they will maintain contact and continue to develop the situation or delay until the appropriate response force appears or the battlefield commander commits the TCF.

TACTICAL COMBAT FORCE

3-27. When the MP response force encounters or engages threats beyond its ability to defeat, it immediately notifies the higher headquarters. The battlefield commander will then evaluate the situation and commit the TCF to defeat the Level III threat. The TCF is normally a combined-arms organization tailored by the corps or division G3, based on METT-TC. The TCF normally receives fire, aviation, or other support needed to fight and defeat the threat. Once the TCF is identified and before it is committed to battle, it will conduct direct coordination with the MP or other response forces regarding the exchange of reconnaissance information, battle-handoff procedures, and contingency plans for TCF operations. Once the TCF is committed, the MP unit performing as the response force becomes OPCON to the TCF commander.

MP AS THE TCF

3-28. The MP brigade or battalion is capable of conducting TCF missions only when properly augmented. MP augmentation may be in the form of fires, small combat units, aviation assets, or CAS. The specific type of augmentation is METT-TC dependent. Additionally, the MP C^2 headquarters must receive the respective liaison officers to ensure that augmentation forces are synchronized and employed according to their capabilities. The MP commander's situational awareness and battlefield visualization are key elements to TCF operations. Once designated as the TCF, the MP unit commander establishes liaison with the appropriate rear CP to obtain—

- The current rear-area IPB.
- The friendly unit disposition.
- Defense plans.
- Priorities for protection.
- The fire-support plan.

3-29. Based on the above information, the MP commander conducts his own IPB and develops a concept of operations. He then forwards it to the appropriate higher echelon for coordination and approval.

3-30. The MP's ability to employ organic MP assets as part of the TCF is limited by the following factors:

- **MP availability.** Normally, all MP assets available are committed at all times. The specific function and scope that MP units perform during the operation are determined by the Army commander's needs, the intensity of the conflict, and the availability of MP resources. The commander, with advice from the MP leader, must decide which MP operations must be scaled back, delayed, or shifted before the MP unit can be designated as part of the TCF.
- **MP dispersion.** MP units are normally displaced over a large geographical area. Technological capabilities and mobility allow them to operate over great distances. In today's battlefield, a typical MP company employment covers between 1,000 and 1,200 square kilometers and performs numerous missions in support of all five functions. The distance between elements, the reprioritization and movement of other MP units, the difficult terrain, poor roads, and bad weather may slow down the MP's commitment as a TCF.

Chapter 4

Military Police Functions

Military police support the Army commander's mission to win the battle. They help the commander shape the battlefield so that he can conduct decisive operations to destroy enemy forces, large or small, wherever and whenever the Army is sent to war.

MAJ(P) Anthony Cruz

The MP Corps supports shaping and sustainment operations while performing its five basic functions as a flexible, economy-of-force organization. Through these functions, MP units are able to provide the commander with an array of CS operations across the full spectrum of military operations (see Table 4-1).

Table 4-1. MP Functions

	Functions				
	MMS	**AS**	**PIO**	**L&O**	**I/R**
Subtasks	Support to river-crossing and breaching operations and passage of lines Straggler and dislocated-civilian control Route R&S MSR regulation enforcement	Recon operations ADC Base/air-base defense Response-force/ TCF operations Critical site, asset, and HRP security Force protection/ physical security Antiterrorism	IPB support PIAP Active and passive roles Information collection and dissemination Joint, interagency, and multinational coordination	Law enforcement Criminal investigations US Customs operations Related L&O training	EPW/CI handling US military prisoner handling Populace and resource control Dislocated civilians
NOTE: Subtasks not all-inclusive.					

OVERVIEW

4-1. MP assets are limited. Specific functions are performed at any given time and are determined by the supported commander's need, the intensity of the conflict, and the availability of MP resources. The supported commander, through the command's PM, sets the priorities for MP operations.

4-2. The PM (based on METT-TC and the available assets) continuously evaluates the trade-off between the MP support that the commander requires and the MP support that can be provided. To meet the priorities set by the commander's tactical plan, the PM recommends the allocation and employment of MP assets for MP combat, CS, and CSS operations.

MANEUVER AND MOBILITY SUPPORT

4-3. The MMS function involves numerous measures and actions necessary to support the commander's freedom of movement in his AOR. The MP expedite the forward and lateral movement of combat resources and ensure that commanders get forces, supplies, and equipment when and where they are needed. This is particularly important in the modern battlefield where there is a greater geographical dispersal of forces and lengthened LOC.

4-4. The MP maintain the security and viability of the strategic and tactical LOC to ensure that the commander can deploy and employ his forces. The MP support the commander and help expedite military traffic by operating traffic-control posts (TCPs), defilades, or mobile patrols; erecting route signs on MSRs or alternate supply routes (ASRs); or conducting a reconnaissance for bypassed or additional routes. The MP move all units quickly and smoothly with the least amount of interference possible.

4-5. As part of the MMS function, the MP support river-crossing operations, breaching operations, and a passage of lines. They also provide straggler control, dislocated-civilian control, route reconnaissance and surveillance (R&S), and MSR regulation enforcement.

RIVER-CROSSING OPERATIONS

4-6. US forces conduct river-crossing operations to move a large force across a river obstacle with a minimum loss of momentum. The MP play a vital role by assisting the force commander in crossing the river as quickly and efficiently as possible. The crossing is usually planned and conducted by the headquarters directing the crossing. As such, a division crossing operation is conducted by a corps. Whether a brigade or division is crossing, the division MP company may also cross to provide uninterrupted support to the division. In these instances, there is a total reliance on the corps MP to support the crossing. The same is true for breaching operations and a passage of lines.

4-7. MP support for river-crossing operations reduces the crossing time and promotes the efficient movement of vehicles. It reduces congestion, speeds the crossing, and enables the maneuver commander to continue his momentum toward his primary objective. The MP establishes staging and holding areas and TCPs to control movement to and from these areas (according to the traffic-control plan). The MP may be called on to provide security for crossing forces at the crossing sites. In most cases, the MP TCPs and engineer regulation points (ERPs) are located on both sides of the river to improve communications and coordination between the units.

4-8. MP employment for river crossing is influenced by METT-TC. The number and placement of MP assets supporting a river-crossing operation varies with the size of the crossing force, the direction of the crossing (forward or retrograde), and the degree of enemy resistance expected or encountered.

4-9. The MP operating inside the crossing areas are OPCON to the crossing-area commander for the duration of the operation. The MP operating outside of the crossing area are under the command of their appropriate echelon commander.

4-10. The main thrust of MP support to river-crossing operations is within the immediate river-crossing site. The MP direct units to their proper locations (such as holding areas and staging areas) and ensure that units move through the area within the time listed on the movement schedule. This is a highly critical aspect of river crossing because the number of crossing sites is limited. MP assets are placed where they can stress MMS operations on MSRs leading into the crossing area.

4-11. The MP also provide AS to allow crossing forces to cross the river without losing momentum or forces. On both near and far sides, the MP are used to recon the crossing unit's flanks and rear to enhance security (see FM 19-4).

BREACHING OPERATIONS

4-12. Breaching operations are conducted to allow forces to maneuver through obstacles. Obstacle breaching is the employment of a combination of tactics and techniques to advance an attacking force to the farside of an obstacle that may be covered by fire. It is perhaps the most difficult combat task a force can encounter. Breaching operations begin when friendly forces detect an obstacle, and they end when the battle handover has occurred between the follow-on forces and a unit conducting the breaching operation (see FM 90-13-1).

4-13. The MP support breaching operations in numerous ways. MP assets are employed based on METT-TC, the available resources, and the commander's priorities. As a minimum, MP support may include, but is not limited to—

- Establishing TCPs along routes leading to or departing from the breaching site.
- Establishing holding areas.
- Establishing TCPs at the breaching site.
- Assisting engineers with temporary route signs.
- Establishing straggler-control operations.
- Conducting AS operations.

4-14. The most critical MP support is provided at the breaching site. The MP provide the commander with a means to control traffic flow to appropriate lanes. When multiple lanes branch off of a single far-recognition marker, the MP assist in directing the formation through various lanes. They also assist in modifying the traffic flow when lanes have been closed for maintenance or expansion. The MP conduct close coordination with the crossing-force commander and the TF commander executing the breaching operation. The MP enable the commander to make last-minute changes in traffic flow, thereby giving him increased flexibility to react to the enemy situation.

PASSAGE OF LINES

4-15. A passage of lines is an operation in which a force moves forward or rearward through another force's combat positions with the intention of moving into or out of contact with the enemy. The passage of lines is a high-risk military operation that requires close coordination between the passing unit, the stationary unit, and the MP providing the support.

4-16. The MP help reduce confusion and congestion during a passage of lines. They provide security in areas surrounding passage points and passage lanes to ensure that the passing unit has priority for using routes to and through the areas. The headquarters directing the operations sets the route's priority. The MP can support a forward, rearward, or lateral passage of lines. Before the actual operation, the MP in the AO conduct an area or zone reconnaissance to become familiar with the routes to, through, and beyond the area of passage. This enables the MP to extend the commander's C^2 by providing directions at passage points and by guiding the units through the passing lanes. Maintaining unit integrity and reducing incidents of stragglers is vital to maintaining the passing unit's momentum in a forward passage of lines. The MP perform aggressive straggler- and dislocated-civilian-control operations to prevent possible infiltration of the enemy.

4-17. A passage of lines is usually planned and coordinated by the headquarters directing the passage. A division's passage of lines is planned and coordinated by the corps headquarters. The detailed plans are made and coordinated between the divisions involved. Close coordination between the division and corps PMs is essential. An MP unit may be the unit involved in passing through the lines of another unit. When conducting a delay of a Level II threat, the MP are likely to conduct a passage of lines with the TCF. To avoid fratricide, close coordination between the MP response-force commander and the TCF is imperative (see FM 19-4).

STRAGGLER CONTROL

4-18. Mobile patrols, TCPs, and checkpoint teams return stragglers to military control as part of their operations. Most stragglers are simply persons who become separated from their command by events in the CZ or while moving through the COMMZ. If a straggler is ill, wounded, or in shock, an MP must give him first aid and, if needed, call for medical evacuation (MEDEVAC). If a straggler is uninjured, an MP directs him to his parent unit or to a replacement unit (as command policies dictate). The MP ensure that stragglers attempting to avoid return to their units are escorted back to their command (as a minimum).

4-19. The MP set up special posts for straggler control following NBC attacks or major enemy breakthroughs that result in large numbers of lost, dazed, and confused military personnel. Mobile MP teams operate between posts, and they also direct or collect stragglers. Straggler collection points may be needed if many stragglers are present in a combat theater. If allied forces are present in the theater, each nation establishes a collection point for its own personnel. MP teams are aware of each allied location and are prepared to assist allied soldiers in returning to their respective command. The MP use available transportation assets to transfer stragglers from TCPs and checkpoints to a straggler collection point. At the collection point, they are screened and sorted for removal to a medical facility or returned to their units to reconstitute the tactical commander's combat force.

4-20. The MP report information about stragglers with whom they come in contact. This information is compiled by the MP headquarters and forwarded through appropriate channels to the higher command. Information given by stragglers that is of immediate tactical value is reported without delay.

DISLOCATED-CIVILIAN CONTROL

4-21. The MP expediting traffic on MSRs may encounter dislocated civilians that could hinder military traffic. The MP assist and divert dislocated civilians from MSRs and other areas to I/R facilities. They may also deny the movement of civilians whose location, direction of movement, or actions may be a threat to themselves or to the combat operation. The HN government is responsible for identifying routes for the safe movement of dislocated civilians out of an AO. If needed, the MP assist the civil-affairs unit and the HN in redirecting dislocated civilians to alternate routes established by the HN government.

4-22. The US forces do not assume control of dislocated civilians unless requested to do so by the HN or unless operating in an environment with a hostile government. When the senior US commander assumes responsibility, the MP coordinate with civil affairs to set up TCPs at critical points along the route to direct dislocated civilians to secondary roadways and areas not used by military forces. As required, MWD teams may be used as a show of force or as a deterrent to assist with uncooperative personnel.

ROUTE R&S

4-23. The MP conduct hasty and deliberate route reconnaissances to obtain information on a route and nearby terrain from which the enemy can influence troop movement. A route reconnaissance focuses on continually monitoring the condition of MSRs, ASRs, and other areas. MP patrols look for restricting terrain, effects of weather on the route, damage to the route, NBC contamination, and enemy presence or absence. When enemy activity is spotted, the MP report it, maintain surveillance, and develop the situation. To gather information for proposed traffic plans, they look at the type and number of available routes; and they check load classifications, route widths, obstructions, and restrictions.

MSR REGULATION ENFORCEMENT

4-24. The MP undertake MSR regulation enforcement to keep the routes free for DSS operations. MP units support the command's MSR regulation measures as stated in the traffic-regulation plan (TRP). The TRP contains specific measures to ensure the smooth and efficient use of the road network. It assigns military route numbers and names, the direction of travel, highway regulation points, and preplanned MP TCPs. Most importantly to the MP, it gives the route's control classification. The MP ensure that classified routes are used only by authorized traffic. Vehicles traveling on roads too narrow for their passage or on roads unable to support their weight can obstruct the route.

4-25. To expedite traffic on MSRs, the MP operate special circulation control measures such as—

- Temporary route signing.
- Static posts such as TCPs, roadblocks, checkpoints, holding areas, and defilades at critical points.
- Mobile teams patrolling between static posts and monitoring traffic and road conditions.

4-26. They also gather information on friendly and enemy activities and help stranded vehicles and crews. The MP also place temporary route signs to warn of hazards or to guide drivers unfamiliar with the route. Using these measures, the MP exercise jurisdiction over the road network in the AO and coordinate with the HN (whenever possible) to expedite movement on MSRs.

AREA SECURITY

4-27. The MP perform the AS function to protect the force and to enhance the freedom of units to conduct their assigned missions. The MP who provide AS play a key role in supporting forces in rear-area and sustainment operations. The MP are a response force that delays and defeats enemy attempts to disrupt or demoralize military operations in the AO. The MP's mobility makes it possible for them to detect the threat as they aggressively patrol the AO, MSRs, key terrain, and critical assets. The MP's organic communications enable them to advise the appropriate headquarters, bases, base clusters, and moving units of impending enemy activities. With organic firepower, the MP are capable of engaging in decisive operations against a Level II threat and delaying (shaping) a Level III threat until commitment of the TCF.

4-28. Throughout all aspects of the AS function, the MP perform counteractions to protect the force and to prevent or defeat a Level II threat operating within the MP's AO. MP countermeasures may include implementing vulnerability assessments, developing procedures to detect terrorist actions before they occur, hardening likely targets, and conducting offensive operations to destroy the enemy. The MP use checkpoints and roadblocks to control the movement of vehicles, personnel, and materiel and to prevent illegal actions that may aid the enemy. The use of these control measures serves as a deterrence to terrorist activities, saboteurs, and other threats. However, at the same time, checkpoints and roadblocks expose the MP to these potential threats. To counter this fact, the MP may upgrade or harden vehicles and defensive positions.

4-29. The MP provide combat power to protect the C^2 headquarters, equipment, and services essential for mission success. The MP provide the battlefield commander with a light, mobile fighting force that can move, shoot, and communicate against any threat. Major subtasks associated with the AS function include reconnaissance operations; area damage control (ADC); base/air-base defense; response-force operations; and critical site, asset, and high-risk personnel (HRP) security.

RECONNAISSANCE OPERATIONS

4-30. As part of their AS mission, the MP serve as the eyes and ears of the battlefield commander by seeking out the enemy and reporting information obtained by recon patrols. The MP conduct area and zone reconnaissances, screening, surveillance, and countersurveillance to gain information to help guard against unexpected enemy attacks in the AO. The MP monitor likely avenues of approach and potential LZs and DZs. They become familiar with towns and other populated areas, ridgelines, woods, and other terrain features from which the enemy can influence movements along road networks. The MP pay close attention to areas near facilities designated critical by the commander. These areas include key MSR bridges and tunnels, depots, terminals, logistics-support bases, ammunition supply points (ASPs), communications centers/nodes, and C^2 headquarters. The MWD teams provide explosive detection and personnel detection/tracking capabilities that enhance reconnaissance operations (especially in urban terrain).

AREA DAMAGE CONTROL

4-31. MP units take measures to support ADC before, during, and after hostile actions or natural and man-made disasters. The ADC actions integrate CS and CSS functions for many units. Engineers, medical personnel, and Army aviators work closely to ensure quick relief operations. The MP provide MSR regulation enforcement, refugee control, and some local security when required. As with reconnaissance operations, the MP may use MWD explosive-and personnel-detection capabilities to augment all MP missions in rear-area and sustainment operations.

BASE/AIR-BASE DEFENSE

4-32. The MP are the base and base-cluster commanders' links for detection, early warning, and employment against enemy attacks. The information gathered is dispersed throughout the rear area to help apprise the commander of enemy activities near bases. Base defense is the cornerstone of rear-area security. When the threat exceeds the base/base-cluster capability, the base/base-cluster commander requests MP assistance through the appropriate C^2 element.

4-33. Air-base defense requires special MP coordination with the US Air Force (USAF). The MP treat air bases like any other base or base cluster. A USAF air base may house the base-cluster commander, or it may be a cluster by itself. The MP are responsible for the air base's external defense. Its internal defense is primarily the responsibility of the Air Force's security forces. The security force provides in-depth defense for weapons, weapons systems, command centers, personnel, and other priority resources established by the base commander.

4-34. The security force is trained and equipped to detect, delay, and deny Level I and II threats. If a Level III threat is present, the security force is tasked with delaying actions; however, the HN, a sister service, or other support must be employed to defeat this threat. If the security force requires assistance to defeat a Level II threat, it may rely on MP response forces or another response force to assist in the defense. If available, the MP response force will react to the air-base defense, just as it would for any other base or base cluster within the MP's AO. However, the key to successful MP employment depends on the critical exchange of information before and during the MP employment. Good communications, an understanding of the defense plan, and liaison operations are vital in preventing responding forces from entering a situation that could result in fratricide.

RESPONSE-FORCE OPERATIONS

4-35. The MP are the base and base-cluster commanders' response force against enemy attacks in rear-area or sustainment operations. The MP gather information about the enemy while performing missions throughout the AO. This information provides commanders with enemy activity near bases. When needed, the MP provide a mobile response force to respond to bases under attack and to destroy the enemy. A base commander's defense plan is the cornerstone for protecting rear-area and sustainment operations. The base commander is responsible for defeating all Level I threats. When this threat exceeds his capabilities, he requests MP support. The MP located near bases or patrolling or conducting AS operations consolidate their forces, respond as quickly as possible, and conduct combat operations to destroy the enemy. If needed, the MP conduct a battle handover to the TCF.

4-36. MP forces performing as a response force are capable of conducting the following offensive operations:

- A movement to contact.
- A hasty ambush.
- A hasty attack.
- A delay.
- A call for fire.
- A repel attack against critical sites.
- A defense of critical sites.

4-37. To conduct these missions, the MP consolidate into squads or platoons to delay, defeat, or defend against the threat. See FMs 71-3 and 71-100 for more information on battle-handover operations.

CRITICAL SITE, ASSET, AND HRP SECURITY

4-38. The MP perform their AS function across the entire designated AO. When the MP provide security around a critical site or asset, they usually provide a mobile security screen, taking advantage of its weapons and communications platforms. This standoff protection detects and defeats the threat before it can move within direct-fire range of the facilities. The MP may be tasked to provide detail security to key facilities, assets, and personnel.

4-39. The MP provide security to major CPs and other facilities within the AO. Their employment maximizes mobility, lethality, and communications capabilities as a security screen. They may be required to establish local AS measures (such as checkpoints and listening/observation posts) to further protect these facilities. The MP provide internal access-control points to critical facilities, and they act as a response force. When the critical CP relocates, the MP provide in-transit security. Other types of critical site security include ASPs; deep-water ports; petroleum, oil, and lubricants (POL) terminals and pipelines; trains and railways; and air bases.

4-40. The MP may provide convoy security for top-priority units transporting especially critical supplies to combat forces. MP assets should be employed primarily on aggressive patrolling, route, area, and zone reconnaissance measures that would create a safe and secure environment for all types of vehicular and unit movement.

4-41. The MP and the CID provide protective services to designated key personnel by providing access control to restricted areas within CPs, providing in-transit security, or providing static security measures around the clock. The MP coordinate with the CID when close-in protection of key personnel is needed. The MP and the CID also provide training for personal-protection countermeasures. The MWD teams may be employed to enhance MP and CID detection and protection capabilities.

INTERNMENT AND RESETTLEMENT

4-42. The Army is the Department of Defense's (DOD's) executive agent for all EPW/CI operations. Additionally, the Army is DOD's executive agent for long-term confinement of US military prisoners. Within the Army and through the combatant commander, the MP are tasked with coordinating shelter, protection, accountability, and sustainment for EPWs/CIs. The I/R function addresses MP roles when dealing with EPWs/CIs, dislocated civilians, and US military prisoners.

4-43. The I/R function is of humane as well as tactical importance. In any conflict involving US forces, safe and humane treatment of EPWs/CIs is required by international law. Military actions on the modern battlefield will result in many EPWs/CIs. Entire units of enemy forces, separated and disorganized by the shock of intensive combat, may be captured. This can place a tremendous challenge on tactical forces and can significantly reduce the capturing unit's combat effectiveness. The MP support the battlefield commander by relieving him of the problem of handling EPWs/CIs with combat forces. The MP perform their I/R function of collecting, evacuating, and securing EPWs throughout the AO. In this process, the MP coordinate with MI to collect information that may be used in current or future operations.

4-44. Although the CS MP unit initially handles EPWs/CIs, modular MP (I/R) battalions with assigned MP guard companies and supporting MWD teams are equipped and trained to handle this mission for the long term. A properly configured modular MP (I/R) battalion can support, safeguard, account for, guard, and provide humane treatment for up to 4,000 EPWs/CIs; 8,000 dislocated civilians; or 1,500 US military prisoners.

EPW/CI HANDLING

4-45. The MP are tasked with collecting EPWs/CIs from combat units as far forward as possible. The MP operate collection points and holding areas to temporarily secure EPWs/CIs until they can be evacuated to the next higher echelon's holding area. The MP escort-guard company assigned to the MP brigade (I/R) evacuate the EPWs/CIs from the corps's holding area to the COMMZ's internment facilities. The MP safeguard and maintain accountability, protect, and provide humane treatment for all personnel under their care.

4-46. In a mature TO, MP (I/R) units process EPWs/CIs collected by MP teams and other units in the CZ. MP guard companies assigned to the MP (I/R) units guard EPWs/CIs at designated camps (see FM 19-40).

POPULACE AND RESOURCE CONTROL

4-47. Populace and resource control (PRC) denies adversaries or insurgents access to the general population and resources. The MP supports civil-affairs personnel and the tactical commander in planning and conducting PRC programs employed during all military operations. These programs may consist of curfews, movement restrictions, resettling dislocated civilians, licensing, ration control, regulation enforcement, amnesty programs, inspecting facilities, and guarding humanitarian-assistance distributions. The MP also direct dislocated civilians to resettlement camps where they are cared for while NGOs work to coordinate their relocation.

4-48. The MP's security capability, acceptability, and interface with the populace make them suitable as the primary forces of choice in these operations. The MP I/R units are specifically designed to fill this need (see FM 41-10).

US MILITARY PRISONERS CONFINEMENT

4-49. The MP detain, sustain, protect, and evacuate US military prisoners. Whenever possible, soldiers awaiting trial remain in their units, unless reasonable grounds exist to believe that they will not appear at the trial, the pretrial hearing, or the investigation or that they will engage in serious criminal misconduct. Under either of these two pretrial confinement instances, the commander must also reasonably believe that a less severe form of restraint (such as conditions of liberty, restriction in lieu of apprehension, or apprehension) are inadequate. When these circumstances exist and other legal requirements are met, US military personnel may be placed in pretrial confinement under the MP's direct control. Convicted military prisoners are moved as soon as possible to confinement facilities outside of the AO.

4-50. MP confinement operations parallel (but are separate from) the MP's EPW/CI operations. No member of the US armed forces may be placed in confinement in immediate association with an EPW or other foreign nationals who are not members of the US armed forces. A confinement facility is maintained within the TO only if distance or the lack of transportation requires such a facility. When military prisoners are retained in theater, temporary field detention facilities may be established in the CZ and a field confinement facility in the COMMZ (see FM 19-40 and AR 190-47).

LAW AND ORDER

4-51. The L&O function consists of those measures necessary to enforce laws, directives, and punitive regulations. The MP's L&O function extends the battlefield commander's C^2. The MP, in close coordination with the CID, work to suppress the chance for criminal behavior throughout the AO. By coordinating and maintaining liaison with other DOD, HN, joint, and multinational agencies, the MP at all levels coordinate actions to remove conditions that may promote crime or that have the potential to affect the combat force. Crime-prevention measures and selective enforcement measures are also performed as part of other functions. For example, the MP investigate traffic accidents and regulate traffic as part of the MMS function. The L&O function includes major areas such as law enforcement, criminal investigations, and support to US Customs operations. The primary units conducting L&O are the L&O detachments, customs teams, and CID units. Both the MWD team and the MP company (CS) also support the L&O function.

LAW ENFORCEMENT

4-52. Law-enforcement operations assist the battlefield commander in preserving his force. The MP dedicate assets to conduct law-enforcement operations based on the commander's needs. Since MP L&O and CID assets may be limited during the initial stages of any operation, the PM depends on the supported commander's development of an effective crime-prevention program and uses established investigative tools (such as inquiries and AR 15-6 investigations) to enforce rules. This will allow the PM to employ limited MP assets to perform other functions. Law-enforcement operations include responding to civil disturbances, conducting raids, investigating traffic accidents, conducting vehicle searches, supporting the commander's force-protection program, and providing support to HN and civil-enforcement agencies. Law enforcement also includes employing special-reaction teams (SRTs), marksman/observer (M/O) teams, and MWD teams.

4-53. The Army conducts counterdrug-support operations that generally fall within several DOD counterdrug-mission categories. The MP support the Army's role rather than directly participating in civil law-enforcement activities (such as searches, seizures, and arrests). When tasked, the MP provide training to law-enforcement agencies in common soldier skills, physical security, and tactical planning and operations. US Code (USC) Title 18, Posse Comitatus Act, Section 1385 prohibits the use of DOD assets to enforce civilian law (federal and state) except in cases and under circumstances expressly authorized by the Constitution or by an act of Congress.

4-54. In multinational operations, the MP may assist with the creation of multinational police units. Circumstances that may support the establishment of these police forces include existing or negotiated terms of international agreements or security-assistance programs, a multinational operational agreement, or appropriate military directives. The MP provide the capability to train foreign MP and/or reconstitute indigenous constabulary forces as part of stability and support operations. The MP can provide the initial mentoring to these forces and provide temporary, emergency law-enforcement capabilities until the foreign military or civilian police forces are functional (see FM 100-8).

CRIMINAL INVESTIGATIONS

4-55. The MP investigate offenses against US forces or property committed by persons subject to military law. Investigations against minor crimes (such as low-value, personal-property thefts or simple assaults) are normally investigated by the MP's L&O detachment. Investigations against major incidents involving wrongful death, serious bodily injury, and war crimes are referred to special agents of the USACIDC. The USACIDC conducts death investigations in the absence of HN agreements or in conjunction with the HN. The USACIDC special agents are authorized to investigate any alleged criminal conduct in which there is an Army interest unless prohibited by law or higher authority.

4-56. The USACIDC's investigative authority and investigative responsibility outside of the US are determined by international treaty or agreement (including status of forces agreements [SOFAs]), the policies of the HN government, the US ambassador, and AR 195-2. In the absence of such provisions, the following guidelines apply:

- On Army-controlled installations, the USACIDC has the authority to investigate alleged crimes.
- Outside of an Army-controlled installation, the USACIDC may investigate after coordinating with HN authorities.

4-57. In all environments, the USACIDC has the responsibility to investigate all felony crimes involving Army personnel, DA civilians and agencies, and companies working for the Army. The USACIDC investigates war crimes and crimes involving personal and government property affecting the Army's mission (logistics security [LOGSEC]). Other investigations (such as those based on international treaties, SOFAs, and joint investigations with the HN) may be undertaken if requested by the supported commander in support of the overall Army mission. See Chapter 9 for a complete discussion of the USACIDC.

US CUSTOMS OPERATIONS

4-58. The MP support the US Customs Service (USCS), the US Department of Agriculture (USDA), other federal agencies, joint staffs, and commanders who enforce the laws and regulations of the US concerning customs, agriculture, and immigration border clearances. Support to the USCS also includes assistance to federal agencies to eliminate the illegal traffic of controlled substances and other contraband through Army channels. MP support to customs operations are normally performed by specially trained MP customs teams. Although other MP units are not trained in all facets of customs operations, they may assist MP customs teams, the USCS or the USDA, and other federal agencies in the enforcement of applicable laws and regulations. When tasked, the MP/CID supports the investigation of violations of US Codes, DOD or DA regulations, and applicable provisions of SOFAs.

4-59. The MP report violations of customs laws, regulations, inspections, and investigative results to the installation's PM, the supported commander, and affected units. During redeployment from outside the continental US (OCONUS) to CONUS installations, the MP support the USCS or USDA efforts to ensure that personnel, equipment, and materiel meet customs, immigration, and agriculture requirements as stated by all applicable laws and regulations. As with other functions, MWD teams may be employed in support of customs operations for the detection of explosives or narcotics.

> During operation Just Cause, an MP platoon temporarily assumed the customs mission at the main terminal of the Torrijos International Airport located just outside of Panama City. Their mission supported the air evacuation of foreign-national civilians and the redeployment of some of the initial-entry US forces.

POLICE INTELLIGENCE OPERATIONS

4-60. The PIO function supports, enhances, and contributes to the commander's protection program, situational awareness, and battlefield visualization by portraying relevant threat information that may affect his operational and tactical environments. This threat information—whether it is police, criminal, or combat information—is gathered while conducting MP functions. The PIO function—

- Demonstrates the MP's/CID's capability to collect relevant threat information actively or passively.
- Ensures that all information collected while conducting MMS, AS, I/R, and L&O functions continues to be reported through the proper channels so that it can be analyzed by the Intelligence Officer (US Army) (S2) or the Assistant Chief of Staff, G2 (Intelligence) (G2) with support from the appropriate MP echelon.
- Coordinates with USACIDC elements to employ data developed by the USACIDC's programs. These programs include—
  - The Combating Terrorism Program as outlined in AR 525-13 and CIDR 195-1.
  - The Criminal Intelligence Program (CIP).
  - Personal-security vulnerability assessments (PSVAs).
  - A crime threat analysis.
  - Logistics-security threat assessments (LSTAs).

NOTE: The MP/CID must ensure that criminal information is released according to existing controls and restraints.

- Maintains constant liaison and communication with the higher echelon S2/G2; psychological-operations (PSYOP) units; HN police and other law-enforcement agencies; joint, combined, interagency, and multinational forces; the staff judge advocate (SJA); the CMOC; civil-affairs teams; and the force-protection officer.

4-61. The MP brigade commander, the battalion commander, and the PM are responsible for the PIO function. As such, each one must determine the best way to employ the available staff resources to monitor the execution of the PIO function within his command.

IPB PROCESS AND PIO FUNCTION

4-62. The IPB is a systematic, continuous process for analyzing the threat and environment in a specific geographic area. It is designed to support staff estimates and military decision making. Applying the IPB process helps the commander selectively apply and maximize his combat power at critical points in time and space on the battlefield by—

- Determining the threat's likely COA.
- Describing the environment the unit is operating within and the environmental effects on the unit.

4-63. The IPB process consists of the following four steps:

- **Define the battlefield environment.** The S2/G2 identifies the battlefield characteristics that will influence friendly and threat operations, establishes the limits of the area of interest (AOI), and identifies gaps in current intelligence holdings.
- **Describe the battlefield effects.** The S2/G2 evaluates the environmental effects with which both sides must contend. The S2/G2 identifies the limitations and opportunities that the environment offers on the potential operations of friendly and threat forces. This evaluation process focuses on the general capabilities of each force until COAs are developed later in the IPB process. This environmental assessment always includes an examination of terrain and weather, but it may also include discussions of characteristics of geography and infrastructure and their effects on friendly and threat operations.
- **Evaluate the threat.** The S2/G2 and his staff analyze the command's intelligence holdings to determine how the threat normally organizes for combat and conducts operations under similar circumstances. When facing a well-known threat, the S2/G2 can rely on historical databases and threat models. When operating against a new or less known threat, he may need to develop his intelligence databases and threat models concurrently.
- **Determine the threat's COA.** Given what the threat normally prefers to do and the effects of the specific environment in which he is operating, his likely objectives and the COAs available to him are determined. The S2/G2 develops enemy COA models that depict the threat's available COAs. He also prepares event templates and matrices that focus intelligence and identify which COA the enemy will execute (see FM 34-130).

4-64. Although the S2/G2 has the staff responsibility for the command's IPB, he is not the only one who conducts or needs to understand and use the IPB. Every Army commander and staff member must understand and apply the IPB process during the staff planning process. The MP on the battlefield are no exception. The MP employ the IPB process as their first step in developing and implementing the PIO function within their commands.

4-65. During the IPB process, the S2/G2 uses all available databases, intelligence sources/products, and related MI disciplines to analyze the threat and the environment. The PIO function supports this process by providing the S2/G2 with collected police, criminal, and combat information that can directly and significantly contribute to the success of the MI effort. In addition to the combat information, the PIO function provides additional information on possible criminal threats and COAs that may support the S2's/G2's IPB process and that can be used by the commander to upgrade the force-protection posture.

POLICE INFORMATION ASSESSMENT PROCESS

4-66. Like the S2/G2 uses the IPB process to continuously analyze the threat and the environment in a specific geographical area, MP leaders use the PIAP as a tool to continuously collect, organize, interpret, and gain access to police/criminal information in support of the IPB process. Criminal trends may have an impact on the tactical scenario, and the PIAP is a method used to consider this threat and its impact on friendly forces. MP leaders cannot use the PIAP as a substitute for the IPB process—the PIAP complements the IPB (see Appendix E).

PASSIVE MODE PIO

4-67. Every MP conducts the PIO function in a passive mode during their normal day-to-day operations and across the full spectrum of military operations. In the passive mode, PIO are not a stand-alone function; as such, they cannot be separated from other MP functions.

4-68. During the performance of MMS, AS, I/R, and L&O functions, the MP develop and exchange information with other organizations in the AO. The MP obtain information through contact with civilians, NGOs, IHOs, local and HN police, multinational police, and other security forces. If the MP receive, observe, or encounter information (police, criminal, or combat) while performing these functions, they will immediately submit a report to relay information up the chain of command. This report may be in the form of a size, activity, location, unit, time, and equipment (SALUTE) report; a spot report (SPOTREP); or another appropriate report. When the higher echelon (brigade, battalion, or PM) receives this information, it is simultaneously integrated into the ongoing IPB/PIAP and forwarded to the higher echelon S2/G2 (see Figure 4-1, page 4-16).

4-69. If police/criminal information is obtained, the MP—

- Update the previous police/criminal estimates provided to the S2/G2.
- Identify new or potential criminal threats or trends in the AO.
- Consider recommending that the supported commander upgrade the force-protection level.
- Notify adjacent units of the potential criminal threat that may affect their forces.
- Consider reprioritizing MP support to the identified threat area.
- Share information with HN/local police and other agencies.

4-70. If combat information is obtained, the MP—

- Forward the information to the higher headquarters S2/G2.
- Forward the information to the MP chain of command, integrate it into the MP's IPB process and, if necessary, take appropriate action.
- Notify the adjacent unit of the potential threat that may affect their forces.

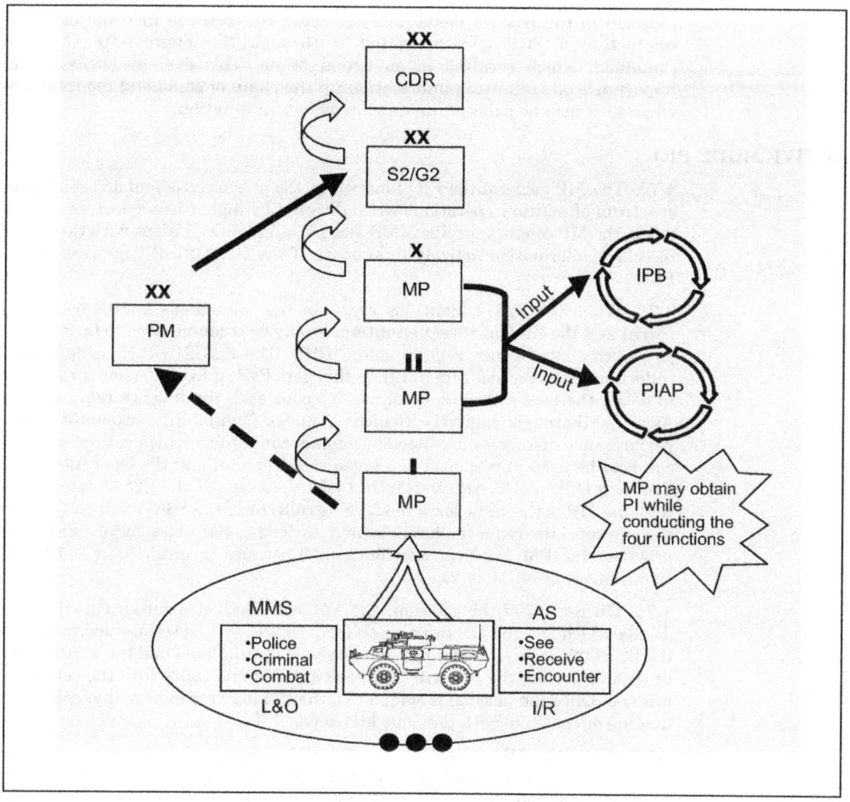

Figure 4-1. PIO—Passive Mode

> During Operation Uphold Democracy, an MP team was conducting a TCP as part of a cordon and search operation in Port-au-Prince, Haiti. While performing the task, two civilians approached the MP team informing them of criminal activity in the neighborhood. During the interview, the MP team prepared a sketch of a house and surrounding areas. The team also obtained information describing the criminals and their weapons. Recognizing that the criminal activity was in fact the actions of a political/mercenary group named FRAP, the MP team radioed the platoon leader and forwarded the field interview to higher headquarters. Two days later, a unit from the 10th Mountain Division raided the house, capturing weapons, ammunition, and equipment.

4-71. The preceding vignette demonstrates the MP performing the PIO function in the passive mode. The MP team received the information while conducting a TCP and submitted it through the appropriate chain of command, which resulted in an action taken. This example stresses the importance of submitting information up the chain of command regardless of whether it may be police, criminal, or combat information.

ACTIVE MODE PIO

4-72. The MP perform the PIO function in the active mode and across the full spectrum of military operations when directed by higher headquarters. In this mode, the MP conduct specific MMS, AS, I/R, and L&O missions with the intent to collect information actively in support of the S2's/G2's IPB process or the PIAP.

4-73. When the S2/G2 identifies a gap in the command's knowledge of the threat and the current threat situation, it may be recommended to be included as priority intelligence requirements (PIR). The S2/G2 will then develop a collection plan to assist him in filling this gap. Part of his collection strategy is to select the best collectors available to cover each intelligence requirement. After a thorough analysis (which includes availability, capability, and performance history), the collection manager identifies which collection asset can best be used in response to a given requirement, and the Operations and Training Officer (US Army) (S3)/G3 tasks the asset. If the S2/G2 determines that the MP is the right force to serve as collectors, the S3/G3 will go through the appropriate request channels and task the PM. The S2/G2 will then provide the PM with a specific guideline and a prioritized collection requirement (see FM 34-2).

4-74. On receipt of the mission, the PM will conduct a mission analysis to decide which specific MP function (MMS, AS, I/R, or L&O) is needed to satisfy the S2's/G2's requirement. Once the analysis is completed and the appropriate function selected, the PM will then task subordinate units with the collection mission. Once the mission is completed, the PM may receive another collection tasking or continue with previous MP tasks.

4-75. When the MP commander or the PM conducts the PIAP, he may also encounter a police/criminal information gap. This gap may become the MP commander's police/criminal information requirements (PCIR). If the gap cannot be filled with available data from the S2/G2, the CID, the HN, and other agencies, the MP commander/PM may task subordinate MP units or request CID assistance to support the collection effort. On receipt of the mission, the MP will then conduct a mission analysis and decide which specific MP function (MMS, AS, I/R, or L&O) is needed to satisfy the PM requirement. Once the mission is completed, the MP may receive another collection tasking or continue with previous MP tasks (see Figure 4-2, page 4-18).

4-76. Since any soldier on the battlefield can report police or criminal information, the MP commander or the PM must constantly coordinate and communicate with the S2/G2, PSYOP, and other agencies to obtain information that could be of MP/CID interest. This constant coordination is a key factor for ensuring that the MP/CID has visibility over the police/criminal information that is reported through non-MP channels.

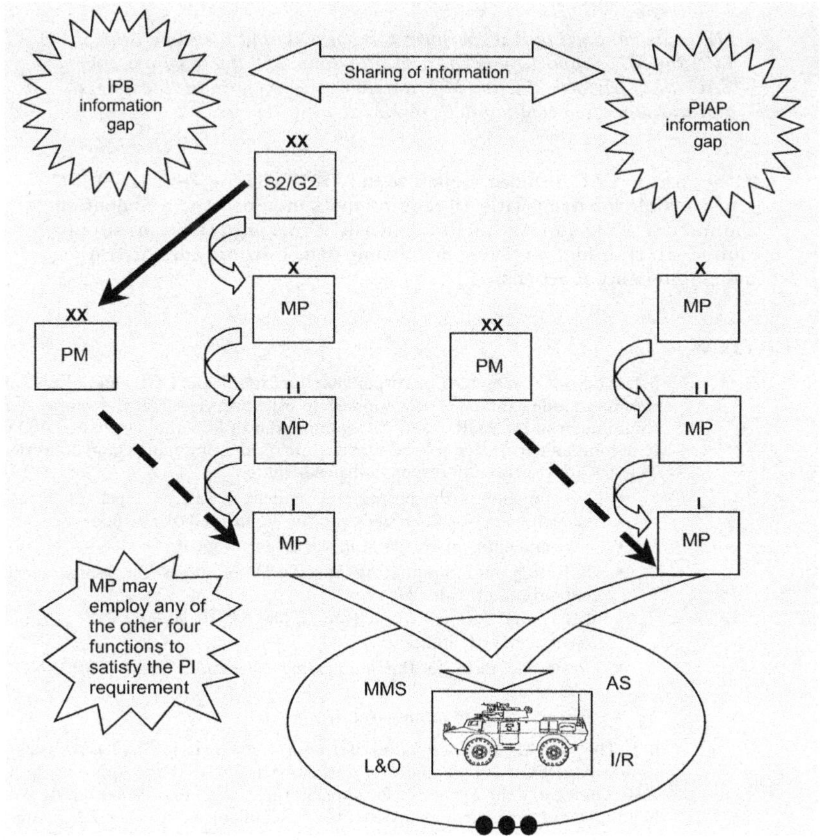

Figure 4-2. PIO—Active Mode

Chapter 5

Military Police Support to Echelons Above Corps

MP units supporting EAC perform combat, CS, and CSS operations. Like the MP supporting corps and divisions, MP units supporting EAC units perform the five MP functions based on available assets and the supported commander's needs.

USAMPS

MP support to EAC includes support to the ASCC and the TSC. The ASCC is responsible for Army Title 10 requirements in support of a combatant commander. This support includes recruitment, organization, supply, equipment, training, servicing, mobilizing, demobilizing, administration, and maintenance functions.

OVERVIEW

5-1. The ASCC may also be responsible for significant DOD- and combatant-commander-designated Army support to other services. As the senior Army commander in the AOR, the ASCC commander tailors and designates ARFOR to accomplish operational-level tasks while conducting major land operations. The ASCC's operational responsibilities include—

- Recommending the proper employment of Army-component forces to the joint-force commander or to the subunified commander.
- Accomplishing operational missions as assigned.
- Selecting and nominating specific Army units for assignment to subordinate theater forces.
- Informing the combatant commander of the Army's CSS effects on operational capabilities.
- Providing data to the supporting operations plans (OPLANs) as requested.
- Ensuring signal interoperability.

5-2. The ASCC provides administrative and logistics (A/L) services to assigned ARFOR and to those of subordinate JFCs. When appropriate, the ASCC delegates the authority for support tasks to a single subordinate Army headquarters. In major operations, the TSC (along with other EAC support commands) would be the ASCC's lead organization for planning, coordinating, executing, or providing required support functions (see FM 100-10).

5-3. The TSC is the senior Army support organization in a theater. Its commander reports to the ASCC or ARFOR commander. The TSC normally operates at the operational level of CSS with links to the strategic and tactical levels. Unity of command is the critical element that the TSC brings to the fight. The TSC is a multifunctional organization that centralizes the command, control, and supervision of support functions at EAC as directed by the ASCC or ARFOR commander. The TSC's mission is to maximize throughput and follow-on sustainment, including all CSS functions, of ARFOR and other designated supported elements. The TSC is capable of synchronizing logistics and other support operations for the ASCC. It provides area support to EAC units in the COMMZ and sustainment support to tactical forces. This support may include supply, procurement, property disposal, maintenance, transportation, field services, health services, civil-military affairs, MP support, engineer support, religious support, finance support, and personnel and administrative services.

5-4. Units and commands requiring support coordinate with the TSC support-operations staff to secure their initial support, to reestablish support, or to resolve support problems. In a fully developed theater, the TSC coordinates with a corps support command (COSCOM) for support of combat forces, although direct coordination with a division support command (DISCOM) is sometimes necessary. The TSC, augmented by a rear operations center, is also responsible for security operations as directed by the ASCC/ARFOR commander (see FM 63-4).

MILITARY POLICE SUPPORT

5-5. MP support to EAC units is provided through an array of multifunctional MP units. The nature of the operation, METT-TC, and the requirements of the supported commander will determine which type of MP unit is appropriate to augment, assign, attach, or place under OPCON to an EAC unit. The types of MP units that support EAC include CS, I/R, CID, and L&O teams (such as MWD or customs teams).

THE MP BRIGADE (CS)

5-6. The MP brigade (CS) provides MMS and AS to extended LOC within the COMMZ. These supply corridors include ports, inland waterways, railways, pipelines, airfields, MSRs, and MSR critical points. The MP support the users of the COMMZ's LOC by aggressively patrolling the area along the LOC. They play an important role in securing rear areas by performing combat operations against the threat. When properly augmented, the MP brigade headquarters may serve as the TSC's/ASCC's TCF headquarters. The MP provide MMS on the COMMZ MSRs leading into the corps's rear area. The MP implement the plans of HN and US staff elements to control the forward movement of combat resources along the LOC.

5-7. If resources are available, the MP brigade (CS) provides escorts to move US noncombatants (if present) from AA points to theater embarkation terminals. Until the MP brigade (I/R) arrives in theater, the CS MP units also perform EPW, confinement, and other operations normally performed by the MP I/R units.

5-8. The organization of an MP brigade (CS) supporting EAC includes the following:

- A brigade headquarters and headquarters company (HHC).
- Up to six MP battalions (each with up to six companies).

- Numerous L&O detachments and MWD teams.

5-9. Additionally, the ASCC's PM or commander may attach or direct OPCON of customs teams to the MP brigade (CS). Battalion and company organization in the MP brigade is the same as that in the corps MP brigade; however, METT-TC determines the number of battalions and companies. The MP brigade (CS) has additional MP companies to provide security for EAC-specific units/activities such as—

- A unified command.
- An ASCC and TSC headquarters.
- LOC seaports, airfields, and railways.
- EAC ammunition storage areas.
- EAC petroleum terminals and pipelines.

5-10. While the corps MP brigade (CS) MWD teams are employed in a GS role, the MP brigade (CS) MWD teams are employed to augment seaport security and to conduct inspections of postal items to detect explosive materiel and narcotics.

5-11. The C^2 within the MP brigade (CS) is consistent with that in any Army brigade. The MP brigade commander works directly for the EAC commander, the battalion commander works for the brigade commander, and the company commander works for the battalion commander. However, this usual C^2 relationship may be altered briefly (based on METT-TC) to enhance the overall EAC combat capability for responding to a Level II threat. For example, MP units operating within an ASG's AO may be under the OPCON of the ASG's rear-area operations center (RAOC), which directly tasks MP units responding to Level II threats. The same is true for placing MP units under the OPCON of the EAC's TCF headquarters for responding to Level III threats. Any conflict in mission priorities is resolved through MP C^2 channels.

5-12. The MP brigade commander is both the MP brigade commander and the EAC's PM. He employs his assets according to METT-TC and the commander's concept of operations. Factors affecting his employment of MP assets include the—

- Nature of the operation (joint, combined, or multinational).
- HN's ability to provide MP-related support (such as port security).
- Custody and location of EPWs/CIs during internment operations until I/R units arrive in theater.
- Number of kilometers of the MSR in relationship to movement-control requirements.
- Number and kinds of critical facilities.
- Number of HRP requiring close-in security.
- HN's ability to control the civilian populace, refugees, and dislocated civilians.
- Supply distribution strategy.
- Risk acceptance and threat in the AO.
- Communications requirements (such as using teams as relays).

5-13. Whenever possible, the MP brigade's AO coincides with the territorial responsibility of the supported command. The MP brigade commander assigns the MP battalion's AO by the above factors as well as by METT-TC. For example, the AOR for an MP battalion may be a large population center of larger geographical areas in which CSS complexes and MSRs are located. But as employment factors and the commander's needs change, so will the MP's AOR. The MP brigade commander must move and tailor his forces to meet the current and projected mission requirements. Unlike many other EAC assets, MP units require 100 percent mobility to shift AOs frequently and rapidly. The following vignette depicts the required MP flexibility to support EAC operations:

> During Operation Desert Shield/Storm, MP units were supporting and moving units throughout their AOs at a such an extraordinary rate that many of them had to relocate their headquarters multiple times just to keep pace with their changing AO.

5-14. Most EAC MP assets are employed along LOC and around areas of high troop concentration. Few EAC MP are dedicated to support fixed commitments (such as ports, air bases, and headquarters [discussed in paragraph 5-8]). When supporting fixed commitments, MP units provide a mobile security screen, and they man static positions when directed or when based on METT-TC. The MP brigade (CS) may have to plan for and actually perform the evacuation and internment of EPWs/CIs and the confinement of US military prisoners until the MP brigade (I/R) arrives.

THE MP BRIGADE (I/R)

5-15. US policy requires that all persons held in Army custody be accorded humane care and treatment from the moment of custody to their final release. The policy applies to detained or interned civilians as well as to EPWs and confined US military personnel. This policy is equally binding on all US troops (see FM 19-40).

5-16. The ASCC or the TSC supports US laws, regulations, policies, and international agreements by providing personnel, administrative, morale, internment, resettlement, and confinement services for the TO. The TSC's MP brigade (I/R) in the TO provides this support. However, since most I/R units are in the reserve components, the initial I/R operations (as mentioned above) may have to be conducted by the MP brigade (CS). Once the I/R unit arrives in the AO, it is responsible for—

- Providing firm but humane control of EPWs/CIs and dislocated civilians.
- Coordinating with HN personnel, military territorial organizations, civilian police authorities, NGOs, private volunteer organizations, and US federal agencies on matters pertaining to I/R operations.
- Performing C^2 operations for all I/R units.
- Controlling, employing, and releasing EPWs/CIs as set forth by the Geneva convention and other international laws and by the UN and other governmental bodies.
- Handling US military prisoners.

5-17. In a mature theater in which large numbers of EPWs are captured, the EPWs' requirements may exceed the capacity of the MP brigade (I/R). In this instance, an I/R command is established. An MP command (I/R) has two or more MP brigades (I/R) and will normally be assigned to the ASCC. When the MP command (I/R) assumes OPCON of the MP brigades (I/R) from the lower echelon, it assumes that echelon's I/R mission.

5-18. If the US decides to transfer captured EPWs/CIs to the HN or to another nation, the US must ensure that the nation is a party to the Geneva convention and is willing and able to comply with the convention. In this case, the number and type of I/R MP units required for processing and retaining EPWs/CIs before the transfer is based on agreements and on METT-TC. Additionally, the MP brigade (I/R) is assigned I/R teams that are located at the processing and transfer points and at the HN or third-country EPW camps. The MP brigade (I/R) liaison team will supervise these dispersed teams to ensure that the HN or the third country provides adequate care and security of US-captured EPWs/CIs and that accountability is maintained according to the Geneva convention.

5-19. The MP brigade (I/R) HHC is the C^2 element for the brigade's assets. It consists of the following elements:

- A brigade command section.
- A company headquarters.
- An Adjutant (US Army) (S1).
- An S2.
- An S3.
- A Supply Officer (US Army) (S4).
- A Civil-Affairs Officer (US Army) (S5).
- A Communications Officer (US Army) (S6).
- Finance and accounting.
- Medical operations.
- Public affairs.
- A unit ministry team.
- An SJA.
- An inspector-general (IG) section.

5-20. Other brigade assets may include the following:

- An MP I/R BLD.
- An I/R information center.
- An MP escort-guard company.
- An MP I/R battalion headquarters.
- MP I/R (EPW/CI) detachments.
- MP I/R (confinement) detachments.
- MP guard companies.
- MWD teams.
- Processing squads, processing liaison teams, camp liaison teams, and evacuation teams (all as required to support EPW transfer or to conduct an out-of-theater evacuation).

5-21. The MP brigade (I/R) subordinate units are employed most often in the COMMZ near CSS facilities and are placed near sea, air, and rail terminals. They receive, process, and intern EPWs/CIs; confine US military prisoners; or assist in the resettlement of refugees or dislocated civilians.

5-22. The MP brigade (I/R) escort-guard company supports the evacuation of EPWs/CIs from the CZ. The company has a company headquarters and four platoons. The escort-guard company requires sufficient mobility to transport MP personnel to the CZ and to operate mobile teams while escorting the EPWs/CIs. The MP go forward to the corps's holding area to take custody of the EPWs/CIs. They may go forward to division collecting points, if distances and resources permit. Using any means of available transportation, the MP ensure that the EPWs/CIs are quickly evacuated to MP battalions (I/R) in the COMMZ. Close coordination with the EAC and corps movement-control centers and the corps MP brigade is required to ensure that transportation assets returning to the COMMZ are employed to evacuate EPWs/CIs from the corps's holding area. Walking wounded EPWs/CIs are evacuated by the same means as other EPWs/CIs, while litter patients are evacuated through medical channels. Guarding EPWs/CIs while in the MEDEVAC channels and during their hospitalization is not an MP mission; therefore, there is not an MP force structure to support this mission. In most instances, the impact of having the MP perform this mission causes trade-offs in missions for which they are responsible.

5-23. The theater MP brigade (I/R) and out-of-theater MP brigade (I/R) subordinate units will evacuate EPWs to internment sites within CONUS (if directed). The theater brigade structure is based on the projected capture rate over time and available out-of-theater transportation assets (frequency and capacity). The out-of-theater brigade structure is based on the total EPW/CI population, the number of internment sites, transportation nodes, and escort requirements. Theater escort-guard MP move the EPWs/CIs to the seaport and aerial port of embarkation (SPOE/APOE). The escort-guard MP assigned to the out-of-theater brigade escort the EPWs/CIs from the theater ports of embarkation (POEs) to the out-of-theater internment sites. The out-of-theater brigade is assigned an I/R evacuation detachment, which is employed at and coordinates the evacuation from the theater POEs, through the out-of-theater ports of debarkation (PODs), to the out-of-theater facilities.

THE MP BATTALION (I/R)

5-24. The MP battalion (I/R) is a modular organization and can be configured to operate internment facilities for EPWs/CIs, confine US military prisoners, or resettle dislocated civilians. When performing EPW/CI internment operations, the MP brigade (I/R) has up to 7 MP battalions (I/R); when augmented with the appropriate number of BLDs, it has up to 21 MP battalions (I/R). The ASCC, the TSC, and the MP brigade (I/R) must consider that the requirement to establish an MP battalion (I/R) internment facility is resource intensive. Therefore, MP I/R units, other supporting units, supplies, and equipment for the EPWs/CIs should arrive in theater ahead of the projected EPW/CI arrival at the internment facilities. Early arrival should be based on the time required to establish fully operational facilities (construct and man) and resupply operations before the EPWs/CIs arrive.

5-25. The MP battalion (I/R) has a command section, a company headquarters, and various staff sections. The staff sections provide the core battalion-level capabilities to conduct internment operations. The modulated design expands as the EPW/CI population increases. The battalion is assigned up to four detachments, two guard companies and, if needed, a processing squad. When fully operational, an MP battalion (I/R) operates an enclosure capable of interning 4,000 EPWs/CIs. The battalion mission centers on eight 500-man compounds. The battalion operates the compounds in close proximity to maximize its resources for the security and internment of the EPWs/CIs. This includes the resources needed to employ EPWs/CIs as a labor force according to the provisions of the Geneva convention.

5-26. Each MP detachment (I/R) (EPW/CI) operates two 500-man compounds and provides augmentation to the battalion staff sections to support 1,000 EPWs/CIs. Each guard company is capable of providing security for 2,000 EPWs/CIs. The guard company has a company headquarters and three platoons. The guard company requires sufficient mobility and communications to support routine battalion missions. While minimum mobility and communications is required to support EPW/CI internment operations, on-site guard personnel must often move considerable distances guarding labor groups performing work projects throughout the COMMZ.

5-27. The MP (EPW/CI) processing squad is capable of processing eight EPWs/CIs per hour and includes interpreters to support the processing. If processing squads are required to augment MP battalions (I/R), the operational requirements will be based on METT-TC.

MP-Battalion Resettlement Operations

5-28. The basic organization used for EPW/CI internment is used for resettlement operations. The primary mission-focus change is from guarding EPWs/CIs to protecting and controlling dislocated civilians. As such, an MP battalion (I/R) with four detachments and two guard companies is capable of supporting 8,000 dislocated civilians. However, the MP battalion (I/R) may require augmentation to conduct L&O operations associated with the resettlement. Augmentation may include the full scope of PM functions (operations, investigations, physical security, MP-station operations, and patrols) and civil affairs.

MP-Battalion Confinement Operations

5-29. When configured with the MP detachment (I/R) (confinement), the MP battalion (I/R) is capable of confining US military prisoners. The MP detachment (I/R) (confinement) provides trained corrections and support personnel required for confinement operations. As with the EPW/CI configuration, the modular confinement structure expands as the US prisoner population increases. With three confinement detachments and three guard companies, the MP battalion (I/R) has a maximum capacity of handling 1,500 US prisoners. Generally, only one MP brigade battalion (I/R) is configured for confinement, but the actual number will depend on the number of US prisoners requiring confinement within a theater. While theater policy for confinement operations remains with the ASCC commander, it is the MP brigade (I/R) that executes the mission.

5-30. When possible, soldiers awaiting trial remain in their units unless reasonable grounds exist to believe that they will not appear at the trial, the pretrial hearing, or the investigation or that they will engage in serious criminal misconduct. Under either of these two pretrial confinement instances, the commander must also reasonably believe that a less severe form of restraint (such as conditions of liberty, restriction in lieu of apprehension, or apprehension) is inadequate. When these circumstances exist and other legal requirements are met, US military personnel may be placed in pretrial confinement under the MP's direct control. Commanders may choose to establish field confinement facilities within their AO. However, corps and division MP companies have the expertise to operate only a field detention facility for a limited period of time. These units cannot operate a confinement facility and have neither the resources nor the capability to operate such a facility on an extended basis. Therefore, all confinement is consolidated in the COMMZ whenever possible.

5-31. All assets of the MP battalion (I/R) (confinement) are employed to detain, confine, sustain, and protect US prisoners. As with the battalions conducting EPW/CI and resettlement operations, the battalion conducting confinement operations is generally located in the rear of the COMMZ, near logistics and transportation support. This allows US prisoners to be moved as quickly as possible from the corps's area to the COMMZ's confinement facility. Movements of US prisoners from the COMMZ to CONUS will be according to DA policy.

Chapter 6

Military Police Support to Corps

Corps MP support their command by performing the MP functions critical to the success of their tactical commander's concept of operations.

USAMPS

Corps are the largest tactical units in the US Army. They are the instruments by which higher echelons conduct operations at the operational level. Higher headquarters tailor corps for the theater and the mission for which they are deployed. They contain organic combat, CS, and CSS capabilities to sustain operations for a considerable period of time. Corps are capable of operating in a joint and multinational environment, providing C^2 for up to five divisions and covering up to 35,000 square kilometers.

OVERVIEW

6-1. MP support to a corps is provided by an MP brigade (CS) assigned to each corps. The MP brigade provides combat, CS, and CSS throughout the corps's AO. However, subordinate MP units are not assigned to subordinate corps units. Instead, the MP brigade commander gives them an AO based on the corps commander's concept of operations. When possible, MP battalion AOs coincide with those of the CSG RAOCs.

6-2. The corps MP provide combat power within the command's rear area. They perform combat operations to counter Level II forces and to support the defeat of Level III forces. When properly augmented, the MP brigade may serve as the corps's TCF. The corps MP also provide a critical link between MP operations in the division and in the COMMZ. The corps MP support division commanders by helping the division MP conduct sustainment operations. The corps MP coordinate with the division MP for mutual support.

THE MILITARY POLICE BRIGADE (COMBAT SUPPORT)

6-3. The MP brigade (CS) supporting a corps contains a brigade headquarters, up to six MP battalions (CS), numerous L&O detachments, and MWD teams. Each MP battalion (CS) has up to six MP companies (CS). As with the EAC's MP brigade (CS), the number of battalions and companies is determined by METT-TC. The corps MP brigade (CS) has additional companies to support each division and to provide security for the corps and COSCOM headquarters and corps ammunition storage areas (see Figure 6-1, page 6-2).

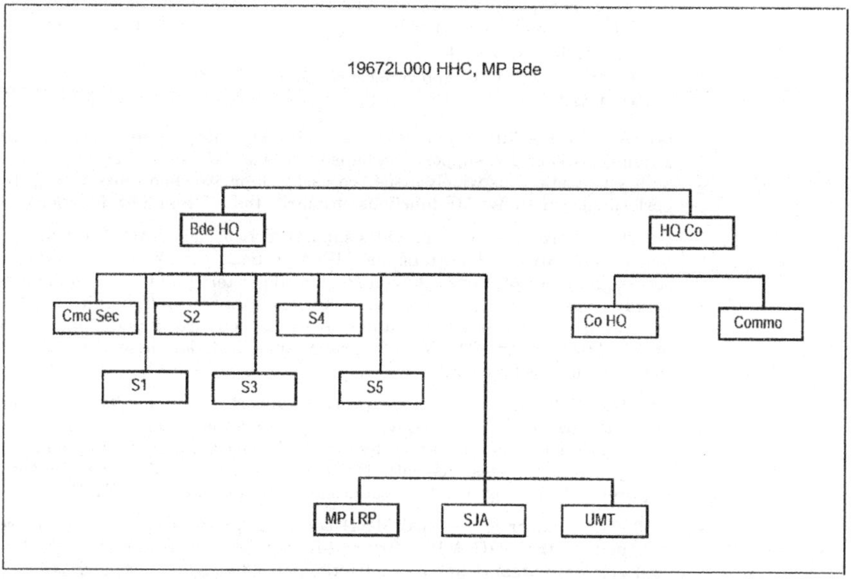

Figure 6-1. Military Police and USACIDC Organizations

6-4. The MP brigade HHC provides C^2 and A/L support to the brigade. The brigade HHC consists of a company headquarters and a brigade headquarters that contains the commander's immediate staff. The staff officers supervise the brigade's major organizational elements, including the—

- S1, S2, S3, S4, and S5.
- MP long-range plans (LRP) section.
- Communications section.
- SJA.
- IG.
- Public affairs.
- Unit ministry team (UMT).

6-5. The MP brigade (CS) command sergeant major (CSM) requires mobility and communications capabilities to execute his duties and responsibilities. The MP LRP section works with the corps G3 plans element, operating out of the corps main CP.

6-6. The MP battalion headquarters and headquarters detachment (HHD) provides C^2 for MP CS companies and any other assigned or attached MP elements. The battalion HHD consists of a detachment headquarters and a battalion headquarters that contains the battalion commander's staff. The staff officers supervise the major organizational elements, including the—

- S1, S2, S3, and S4.
- Communications and support sections.
- UMT.

6-7. As with the MP brigade (CS), the battalion CSM requires mobility and communications. The support section is vital to an MP battalion commander's ability to sustain his widely dispersed assigned or attached units during the performance of all five MP functions, primarily the MMS and AS functions.

6-8. The MP company (CS) provides support to an assigned AO. The company has a company headquarters, an MP operations center, a combat-medic section, and four platoons. The company headquarters provides maintenance, supply, communications, mess, and medical support to the unit. An MP operations center supports the unit's operation, conducting and planning for all five MP functions. The MP operations center includes three MP teams as the company-level response force.

6-9. The MP L&O detachments provide support to an assigned AO. The headquarters provides A/L support. The operation team plans and supervises desk operations, the traffic-accident and MP-investigation teams, and the force-protection teams. As with the EAC MP brigade (CS) detachments, requirements are based on the population supported and on METT-TC.

6-10. The number of assigned MWD teams is significantly less than those assigned to the EAC MP (CS) or (I/R) brigades. A kennel master, five explosive/patrol teams, and four narcotics/patrol teams are normally assigned to the corps MP brigade (CS). They are employed based on METT-TC.

COMMAND AND CONTROL

6-11. The C^2 in a corps MP brigade (CS) is consistent with that of any Army brigade. The MP brigade commander commands the brigade and all attached personnel. Battalion commanders work for the brigade commander. The MP company commanders receive their orders and work for their respective battalion commanders and direct their platoon leaders according to mission requirements.

6-12. When corps MP assets are sent to augment the division PM or the MP company, they are placed under the OPCON of the division PM for as long as the unit is needed in the division area. This command relationship is applicable to company-size organizations and smaller. If an MP battalion (or larger MP unit) augments the division, then the division PM will not exercise OPCON of that MP unit. The MP battalion will be placed OPCON as designated by the division commander but not under the division PM. One L&O detachment is normally attached to the supported division and placed OPCON to the division PM. However, MP leaders at each level establish an MP C^2 relationship based on METT-TC and the supported commander's needs.

EMPLOYMENT

6-13. Each corps MP brigade commander employs his assets according to METT-TC, the needs of the forces operating in his AO, and the priorities of the corps commander. Few MP assets in the corps area are employed to support fixed commitments. Instead, based on a broad consideration of the enemy and friendly situations, the corps MP are employed to support friendly forces engaged in combat, CS, and CSS operations. Since MP forces are dispersed throughout the corps area, the concentration of US forces, the location and vulnerability of critical sites, and the number of kilometers of the MSR to be controlled influence the designation of MP assets in the AO.

6-14. The MP brigade commander establishes his MP concept of operations based on the corps commander's concept of operations. His successful employment of MP assets depends on his foreseeing where the battle will be rather than where the battle is. Based on the rear-area IPB and PIAP, the MP brigade commander allocates and shifts resources to ensure the accomplishment of priority missions. This ensures the continuous support and forward sustainment of combat units and the safety of CSS units operating in the corps rear and sustainment areas.

6-15. To support the MP brigade commander's planning, the MP at brigade headquarters operate from several locations. Most of the staff locates in the vicinity of the corps rear CP where they can interface with the corps staff responsible for planning and executing rear operations. The control element of the brigade headquarters must be located where it can command and control its subordinate units. The brigade commander and his staff decide the best place to locate this element. The planning element of the headquarters locates near the corps main CP where it can interface with the corps commander's coordinating, special, and personal staffs. From there they monitor MP operations, integrate MP support with the corps plans cell for future operations, and learn the enemy situation through the G2's threat analysis almost immediately. The detailed information on rear-area activities and operations provided by the corps staff enhances the accuracy of the MP LRP.

6-16. The brigade S3 section provides the day-to-day planning and execution of an MP mission. The section provides a responsive CP that can relocate frequently throughout the AO. The S3 will normally provide liaison personnel to the corps rear CP, selected COSCOM units, interagencies, or other headquarters (based on mission requirements). When possible, battalion AOs coincide with the AOs of the CSG RAOCs. The MP brigade commander usually tailors battalions' AO boundaries to ensure responsive and flexible support across the corps's AO. He pays particular attention to the LOC behind the most heavily committed division and the critical bases and facilities in that area. He also ensures that the MP are available to respond quickly to combat operations throughout the entire corps rear area or during sustainment operations.

6-17. The MP brigade commander, coordinating with his battalion commanders, locates the MP companies where they can provide combat and CS power throughout the corps's AO. He bases his decision on the—

- Number and composition of urban areas.

- Location of CS and CSS units.
- Location of critical facilities (such as the headquarters, ammunition storage areas, and airfields).
- Expected threat.
- Level and frequency of support needed by the commander.
- Current and projected tactical situation.
- MSR network, including choke points and critical bridges and tunnels.
- Number of supported divisions and requirements.

6-18. The prioritization of MP missions is especially crucial during the early stages of the deployment when it is unlikely that an MP brigade commander will have a full complement of CS companies. Augmenting the division MP company with corps MP assets may not be possible at that time. Until it is, the MP brigade commander must locate corps MP assets to help meet the needs of the division while fulfilling the needs of the corps.

6-19. Like the brigade commander, battalion commanders plan the employment of their companies and platoons using METT-TC. Certain corps needs are constant. One MP company is allocated to provide security for the COSCOM, and one MP company is allocated to provide security for the corps's main CP. One platoon from that company may be used to secure the corps tactical CP or the jump CP. One or more platoons will also help secure the corps's rear CP. The number of MP assets allocated for a corps-level EPW/CI holding area and escort is based on METT-TC. However, a minimum of one platoon is dedicated to operate the corps's EPW/CI holding area and a minimum of one platoon per division is allocated for evacuating EPWs/CIs from division collection points. Additional MP assets may be allocated to provide security for the corps's ammunition storage area and ASPs supporting the divisions. The MP battalion commander places his companies where—

- MP assets support the brigade commander's concept of operations.
- The MP can support troop concentration, bases and base clusters, road networks, and critical areas.
- The MP can aggressively patrol critical terrain and monitor LZs and DZs to detect or deny enemy interference.
- The MP can respond to Level II threats.
- The MP can support the movement of combat resources throughout the AO.
- The MP can remove EPWs/CIs from division collection points.
- The MP can influence stragglers, refugees, and dislocated civilians.

6-20. Battalion commanders may choose to place a company behind the division rear boundary. This can help to ease the coordination between the corps MP and the division MP.

Chapter 7

Military Police Support to Division

Division MP provide essential support to the forward tactical commander. The MP provide a flexible, mobile, and lethal force ready to be employed in combat or CS operations.

USAMPS

The division is a large Army organization that trains and fights as a tactical team. Largely self-sustaining, it is capable of independent operations. The division is a unit of maneuver organized with varying numbers and types of combat, CS, and CSS units. Divisions fall mainly into two categories—heavy or light. The first category includes heavy divisions, their variants, and the Division XXI digitized heavy division. The second category includes the light infantry, airborne, and air-assault divisions. Each division can conduct operations over a wide range of environments. The success of Army operations depends on the success of its divisions.

OVERVIEW

7-1. In corps operations, divisions are normally comprised of three maneuver brigades, each with up to nine maneuver battalions, artillery battalions, aviation battalions, and supporting CS and CSS units. Divisions perform a wide range of tactical missions and are self-sustaining for limited periods of time. The corps augments divisions as the mission requires. All divisions must be able to deploy and conduct offensive, defensive, stability, and support operations. Airborne and air-assault divisions must be able to conduct forced-entry operations. Divisions may be part of a JTF or serve as the ARFOR headquarters.

7-2. MP support to a division is provided by an organic MP company that includes the PM and his staff (assigned to each division) and is augmented by the corps MP brigade. The division MP company provides support in the DRA and in the maneuver brigade rear area. It is fully mobile in order to relocate frequently under short notice. While division MP companies are capable of performing all five MP functions, they must be augmented by corps MP/CID to fully perform all five MP functions throughout the division's battle space. The division PM must receive at least one corps MP company for GS missions in the DRA, one L&O detachment for L&O missions, and one CID detachment for criminal investigations and LOGSEC. In a mature theater (or based on METT-TC), the division PM can expect two corps MP companies. In addition, the division PM may request a corps MP battalion headquarters to provide C^2 to corps MP units operating in the DRA or to perform missions as a TCF. Because the need for MP support exceeds division organic assets (and many times exceeds augmenting corps MP assets), careful planning of MP employment is essential.

DIVISION MILITARY POLICE COMPANY

7-3. The organization of the division MP company is set by specific tables of organization and equipment (TOE) and is designed to support a specific type of division. Regardless of the type of division, the assets in a division MP company enable a PM and his supporting staff to supervise the performance of all five MP functions. The PM coordinates the employment of the MP assets in the division's AO through his PM section. The division MP company contains a headquarters, a PM section, a medic section, and three to six platoons.

COMMAND AND CONTROL

7-4. The C^2 of the division MP company extends downward from the tactical commander to the division PM. The division PM is the division commander's special staff officer. Under normal circumstances, he works for the chief of staff. The division PM has OPCON of the division MP company and MP assets that have been provided by the corps. The division PM plans and employs all of the organic, assigned, or attached MP according to METT-TC. He anticipates the support needs of the division commander and plans the employment to meet his needs. After learning the division commander's concept of operations and estimating the situation, the division PM decides which MP functions are critical to accomplish the commander's mission. He then recommends the allocation of resources, tasks his assets, and establishes the priority of support.

7-5. When the division MP company is augmented by corps MP platoons, the platoons are placed under the direction of the division MP company commander, who assigns an AO to each platoon. When corps MP assets augmenting the division make up a company, the company is attached to the division. The company then comes under the OPCON of the PM, who assigns it an AO. The augmenting corps MP company is usually given an AO from the division rear boundary forward. However, this assignment is flexible and based on the division commander's needs as set forth by the division PM. The four platoons of that corps MP company perform all five MP functions in that AO. Additionally, the corps MP brigade (CS) L&O detachment supporting a division is attached to the division and OPCON to the division PM.

7-6. The division MP company commander directs the employment of company assets. Through his company headquarters, he provides administrative, maintenance, and logistical support to the PM section and to the company's platoons that are dispersed throughout the division's AO. These sections are mobile to support the platoons without delay despite the extended distances that may be separating them. The company commander is also mobile. He travels the extended distances from one platoon area to another when checking on the status of his soldiers and resolving the problems affecting mission accomplishment. The division HHC provides mess support. Personnel administration is handled by the company headquarters and the DISCOM. The number and kind of assets in a division MP company and the configuration of its platoons and squads are determined by the type of division to which the company is assigned.

NOTE: See the consolidated TOE update that is published yearly for complete details (base and objective) for all MP TOEs.

7-7. At the direction of the division commander, the division band may be available to perform its secondary mission of supporting MP operations. When the intensity of the conflict prevents the band from performing its musical mission, the division commander may direct it to augment the MP company. The band, normally employed as a unit under the direction of the bandmaster, is under the OPCON of the MP company commander while augmenting the MP. When tasked to augment MP assets, band members may be employed to augment MP security at the division main CP or to augment perimeter security at the division EPW/CI collection point. The band is released to perform its primary mission as soon as the tactical situation permits.

MILITARY POLICE EMPLOYMENT

7-8. The employment of division MP companies differs somewhat with the type of division to which they are assigned. In heavy divisions, the division MP companies provide GS to the DRA and DS to the division's maneuver brigades. When the platoon is providing DS to a maneuver brigade, the MP platoon leader coordinates all logistical support with the supported brigade. A platoon's DS relationship is seldom interrupted. However, METT-TC may require weighing MP support to the maneuver brigade tasked as the main effort or employing all platoons in a division MP company to function as a unit (such as during division-size river-crossing operations or deliberate breaching operations). In light divisions, the MP provide only GS to the DRA or during sustainment operations. Support exceptions are addressed under MP support to airborne and air-assault divisions, discussed later in this chapter.

7-9. In the division (where flexible support of an austere force is crucial), the division PM must have a clear understanding of situational awareness. To obtain current information for projecting MP needs in the division area, he must be mobile and be able to conduct split-cell operations. The assets available to the PM include the division MP company and at least one corps MP company. Corps augmentation is required for sustained operations and for special operations such as river crossings, dealing with dislocated civilians, and refugee internment or resettlement. The division PM coordinates with the corps PM and the MP brigade or CID commanders for—

- Evacuating and guarding EPWs/CIs from division to corps.
- Providing law-enforcement assistance to HN forces in the division's AO.
- Providing corps augmentation for the division's AO, convoy security, LOC security, AS, R&S, L&O, and other missions. These other missions may include augmentation for security of the division main CP, ASPs, and other critical facilities.
- Integrating, sharing, and exchanging police intelligence between corps and division MP elements.
- Providing CID support.

SUPPORT TO THE HEAVY DIVISION

7-10. The Army's armored and mechanized infantry divisions (normally referred to as heavy divisions) provide mobile, armor-protected firepower that is normally employed for their mobility, survivability, lethality, and psychological effect (shock) on the enemy. These divisions destroy enemy armored and mechanized forces, and they can seize land areas and secure key terrain. Because of their strategic lift requirements, heavy divisions are slow to deploy from home staging bases into an AO. They have high consumption rates of supplies and have limited use in restrictive terrain. These capabilities and limitations are key factors in planning effective and efficient MP support throughout the division's AO.

7-11. In the heavy divisions, the PM section is organized to support split-cell operations at the main and rear CPs. The PM must be mobile to ensure that he is fully aware of the current status of critical MP operations. Therefore, the deputy PM (DPM) locates in one of the PM cells and—

- Handles the section's routine operations.
- Monitors ongoing division operations and MP support.
- Helps provide long-range planning and interface with the primary division staff.
- Forwards PM taskings to the division MP company commander.

7-12. The DPM and the operations sergeant normally set up operations at the division main CP where they can coordinate requirements with the division staff personnel. The company headquarters is initially located near the division rear CP in the division support area (DSA). Once augmentation arrives from the corps, the headquarters relocates with one of its GS platoons operating behind the brigade rear boundary. However, this location could be changed based on METT-TC, the supported commander's needs, and the company commander's idea of where he can exercise better C^2 for his unit.

7-13. The Army of Excellence (AOE) heavy division MP company has six platoons. Three platoons provide support to each maneuver brigade and are designated as DS. The other three platoons are designated as GS platoons. One MP platoon provides security for the division main CP; one provides security for the division's EPW central collection point; and one performs other MP operations within the division rear.

7-14. The GS MP platoons' AOs are configured based on METT-TC and the availability of MP augmentation from the corps. The DS MP platoons' AOs coincide with the supported maneuver brigade's boundary. Each platoon headquarters locates within its brigade's support area or any other area where it can best provide and receive support. To accomplish its mission, each DS platoon requires a minimum of two squads, each with three teams. One squad operates the EPW/CI collection point. The other squads perform MMS and AS operations. All MP platoons are capable of performing all five MP functions. However, performance of these functions is prioritized based on METT-TC and the division commander's concept of operations. The division PM, the company commander, and METT-TC dictate how these platoons should be tasked-organized to accomplish the mission.

SUPPORT TO THE LIGHT INFANTRY DIVISION

7-15. The AOE light infantry division (LID) is one of the most rapidly and strategically deployable divisions. It fights as part of a larger force in war or conducts missions as part of a joint force in MOOTW. Its C^2 structure readily accepts any augmentation forces, permitting task-organizing for any situation. The augmentation required for the division is largely determined by METT-TC. The division's capabilities allow it to exploit the advantages of restricted terrain and limited visibility. It achieves mass through the combined effects of synchronized small-unit operations and fires rather than through the physical concentration of forces on the battlefield. These characteristics are key factors in planning and employing MP assets in support of the LID.

7-16. The LID MP companies are capable of performing all five MP functions. However, their performance of these functions is prioritized based on METT-TC and the division commander's concept of operations. Contrary to the heavy division MP company, the LID MP companies are much smaller. The constrained size of the LID MP companies makes corps MP augmentation crucial to the sustainment of MP operations. Additionally, the LID MP company is the only MP unit with the capability of antiaircraft support through the use of shoulder-fired air-defense weapons.

7-17. The company has three GS platoons to support the division. No platoons are provided to the maneuver brigade. One platoon is normally located in the vicinity of the division main CP so that its resources can help support CP security. Another platoon locates in the DSA and operates the division EPW/CI collection point. The last platoon has an AO configured according to METT-TC and the commander's priority of MP missions. Each GS MP platoon has a headquarters and three squads, each with two teams. The PM section is located in the vicinity of the division main CP. The exact location is based on the current operational status and on METT-TC.

SUPPORT TO THE AIRBORNE DIVISION

7-18. The AOE airborne division can rapidly deploy anywhere in the world to seize and secure vital objectives. It conducts parachute assaults to capture initial lodgments, executes large-scale tactical raids, secures intermediate staging bases (ISBs) or forward operating bases (FOBs) for ground and air operations, and rescues US nationals besieged overseas. It can also serve as a strategic or theater reserve as well as a reinforcement for forward-presence forces. The airborne division can assault deep into the enemy's rear area to secure terrain or to interdict enemy supply and withdrawal routes. It can seize and repair airfields to provide a FOB and airheads for follow-on air-landed forces. It is capable of all other missions assigned to LIDs. The airborne division does not have sufficient armored protection to defeat heavier armored formations at close range. Therefore, engagements with enemy armored formations require special considerations. Antitank weapons in the division compensate for, but do not completely offset, this deficit.

7-19. Airborne divisions conduct operations in two phases—assault and defense. The division accomplishes the assault phase in three stages. First, they deploy; second, they establish the airhead; and third, they extend the logistics base and build their force.

7-20. Like other division MP companies, those assigned to the airborne division are employed to support their division commander's concept of operations. The airborne MP company has four platoons, each having three squads with two teams. Normally, the company headquarters and one of the platoon headquarters colocates with the PM section. The division PM's exact location depends on METT-TC. During the assault phase, the airborne division MP company provides DS to the assault brigade. The remaining platoons provide MMS and AS in the vicinity of the division main CP.

7-21. The nature of airborne operations makes the capture of EPWs likely. Therefore, during the first stage of the assault phase, the priority of MP support is given to EPW operations. After assembling the DZ or LZ, the MP collect EPWs captured during the assault. Combat elements are relieved of EPWs as far forward as possible. In airborne operations, EPWs are held for later movement to a central collection point. During the first stage of the assault, the MP perform limited straggler and refugee control and undertake AS operations, when possible.

7-22. MP support is reevaluated after the airhead is established in the second stage of the assault. The PM takes in consideration personnel and equipment flow, roadways, and security requirements to shift MP support priorities. As the entire operation matures, MP support may expand to include all five MP functions.

SUPPORT TO THE AIR-ASSAULT DIVISION

7-23. The air-assault division executes tactical missions at operational depths to achieve strategic results. It is capable of launching brigade-size air assaults of nearly 4,000 soldiers from either an ISB or a tactical assembly area (TAA). Within 6 hours, this air-assault task force (AATF) can attack an opposing force, occupy and defend key terrain, or establish a FOB (out to 150 kilometers) from which even deeper operations can be executed. Air-assault operations are terrain independent, but they rely on suitable weather and a detailed attrition/assessment of enemy capability—particularly air-defense assets along air corridors and in the objective area. The air assault is preceded by detailed, lethal and nonlethal condition settings, culminating in a comprehensive condition check before execution.

7-24. As the corps AATF unit, the brigade combat team (BCT) task-organizes and habitually trains with both aviation lift and attack (Apache) battalions. An air-assault division MP company has four platoons, each having three squads with two teams. To facilitate operations, air-assault division MP platoons are habitually aligned with each of the three AOE BCTs; however, as with aviation assets, the MP are task-organized to support the main effort's BCT/AATF. When conducting these operations, two MP platoons may be tasked to provide support; one is available to posture and participate in the air assault, and the other conducts AS and MMS operations at the pickup zone (PZ). This second MP platoon may further provide MMS and accompany a ground-assault convoy (GAC) to the objective, or it may remain at the PZ to receive EPWs returning on aircraft from the objective area. The mission of conducting MMS along this vulnerable ground LOC is particularly critical if the AATF objective is a FOB from which further division operations will expand.

7-25. MP participation in the actual air assault competes for heavy lift with FA and air-defense systems, as well as with vital Class V resupply. However, the lethality and versatility of the MP bode well for their employment, and two MP platoons are available to support the brigade air assault as the division's center of gravity.

7-26. When possible, habitually aligned platoons remain with their brigades, and corps assets perform GS missions. However, when no corps assets are available and two division platoons are employed as stated above, the two remaining platoons conduct division EPW collection-point operations and other MP functions based on METT-TC. Normally, the EPW platoon and the MP company headquarters colocate in the DSA. As required (and based on METT-TC), airflow planning includes EPW/CI evacuation from the AATF/ FOB collection point back to the DSA. The PM section operates from the division rear CP to facilitate I/R operations and to coordinate MMS and AS with key logistical staff. Due to potentially extreme distances on the air-assault battlefield, the DPM normally locates with the division main CP to serve as a key G3 battle-staff member and to coordinate PIO with the G2.

SUPPORT TO DIVISION XXI (HEAVY DIVISION)

7-27. Division XXI represents a significant change in the manner in which division operations are conducted. These changes are brought on by information-age capabilities; an increased integration of service components into an effective battle team; more lethal, survivable, and agile systems; and more capable soldiers and leaders. The Division XXI operates in a larger battle space and at a higher tempo than the AOE division. The division is evolutionary in design, but revolutionary in its use of information technology. It improves the Army's deployability while enhancing its ability to dominate in decisive fights. The following are characteristics of the Division XXI operational environment:

- **Multidimensional.** The division will operate in an extended battle space that goes beyond the traditional physical dimension of width, depth, and height. It includes portions of the electromagnetic spectrum; and it extends beyond the physical boundaries of the division through its communication and digital connectivity to other Army, joint, and coalition elements, even reaching back to CONUS.
- **Precise.** Precision operations go beyond a precision strike to include every aspect of military operations from deployment through combat and through redeployment or transition to other operations. Precision in decisive operations is enabled by three emerging capabilities. First, digitization provides soldiers and leaders at each echelon the information required for making decisions. Second, a full suite of strategic, operational, and tactical sensors linked to analytical teams fuses combat information into situational awareness across the battle space. Third, simulations enable Army elements to be tailored and operations to be planned, war-gamed, and rehearsed—yielding precision execution.

- **Nonlinear.** Nonlinear operations do not seek a battle-space grid of close, deep, and rear operations. Instead, the battle space is fluid, changing as METT-TC changes through the duration of mission preparation and execution. Another dimension of this characteristic is the synchronization of near-simultaneous operations to achieve nonlinear effects across the battle space.
- **Distributed.** Division operations are distributed or executed where and when required to achieve decisive effects concentrated at a decisive point. Dispersion empowers subordinates to operate independently within the commander's intent, leading to synergistic effects that exceed the effects of a centralized headquarters.
- **Simultaneous.** The concept of decentralization operations that are multidimensional, precise, distributed, and nonlinear yields the capability to conduct simultaneous operations across the battle space. Simultaneous operations seize the initiative and present the enemy leadership with multiple crises and no effective responses. Rather than a single, concentrated attack, the division executes a series of attacks (lethal and nonlethal) as simultaneously as possible.
- **Integrated.** Division operations are fully integrated with joint, multinational, and nongovernmental partners. Integrated operations enable the Army to leverage the full suite of capabilities that the services bring to the battle space.

7-28. The Division XXI heavy division is one of total integration—not rounding up or down. Reserve-component soldiers assigned to the division are an integral part of the division mission and task accomplishment. The division cannot operate without them. The division is focused on massing effects on the enemy—enabled by a new foundation of information technology and distributed logistics. The division is built around the pattern of "conduct decisive operations," accomplished by either fires or maneuver. The design performs complementary battle-space shaping in concert with the overall corps plan. Its long-range assets are not typically held in reserve nor just employed in the close fight; they are used to best support the corps and division commanders' plans of operations.

7-29. Although the basic structure of the division has not changed—three maneuver brigades, a division artillery, and a DISCOM—the internal structure and operational construction have changed in a near revolutionary way. The C^2 structure and the basic organizational structure are very similar to the divisions discussed above. However, by integrating the capabilities of digital technology and by using enhanced situational awareness, it is better able to execute combat operations.

7-30. The maneuver brigades have their own scouts—a brigade reconnaissance troop (BRT); therefore, they no longer rely exclusively on the battalion scouts. The maneuver battalions are reduced to three companies with a total number of 45 combat platforms. The division artillery has three Multiple-Launch Rocket System (MLRS) batteries per battalion. The cavalry squadron is under the aviation brigade's C^2 but retains and performs the division's traditional cavalry mission. Figure 7-1 depicts the organizational structure of Division XXI.

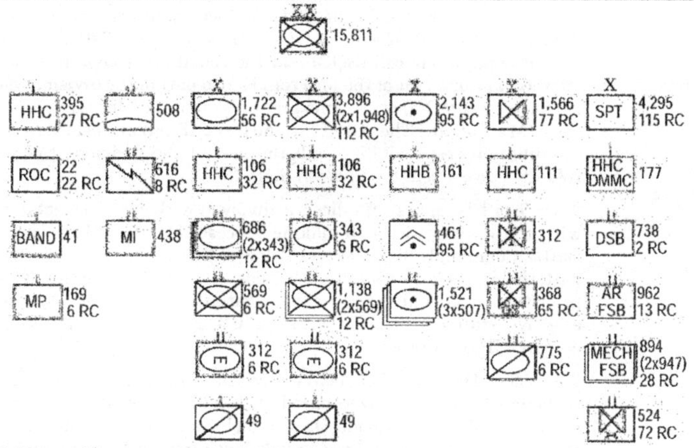

NOTE: Personnel figures are estimates; they may change.

Figure 7-1. Division XXI Structure

7-31. The division's CSS is centralized. Instead of each unit having its own organic CSS personnel, they are now centralized in CSS units. The CSS assets are reorganized and reassigned to the DISCOM. Forward-support companies (FSCs) are habitually associated with each maneuver battalion and forward-support battalions (FSBs) will support each maneuver brigade. This operational structure enables distribution-based rather than supply-based logistics concepts to support the division. The overall logistics footprint for the division is now reduced and the division has organic self-sufficiency to operate unsupported for up to 30 hours.

7-32. A major difference between the AOE fix-forward doctrine and the Division XXI maintenance support is the change in the actual repair site. Division XXI maintenance units directly supporting a given customer will diagnose faults and will replace components forward in the battlefield and will repair rearward. Units will be manned with maintenance personnel capable of performing DS and limited GS maintenance.

7-33. The division operates in a battle space two and a half times larger than an AOE division, and it will usually fight as part of a corps or a JTF. The division conducts offensive, defensive, stability, and support operations in cooperation with other elements of the corps/JTF to gain tactical or operational results significant to the joint-force commander's campaign goals.

7-34. Although the division is a heavy division, MP support is somewhat different from that of an AOE heavy division. The PM (based on METT-TC) establishes operations where MP assets can best be commanded. At the same time, overall MP operations can be synchronized to support operations within the division rear and the division commander's intent can be effectively supported. To this end, the PM conducts split-cell operations within the division main CP. Normally, the PM section is distributed between the security-operations cell (SOC) and the mobility/survivability (M/S) cell. The DPM (who works out of the M/S cell) is also part of the division planning team. He ensures that MP operations are integrated into division plans according to the PM's intent. The DPM attends coordination meetings and assumes coordination responsibility with the rest of the division staff when the PM relocates away from the division main CP.

7-35. The PM, as a staff officer, is involved with every aspect of planning and with the military decision-making process (MDMP). He uses the Force XXI Battle Command—Brigade and Below (FBCB2) C^2 system to understand the battlefield. He understands the mission and the commander's intent and priorities. The PM has complete situational awareness, establishes an effective staff relationship with the G3, and recommends the best way to employ MP resources.

7-36. The Division XXI heavy division MP company has five platoons. Although tasked-organized based on METT-TC and the commander's needs, three MP platoons provide DS for each maneuver brigade. The DS platoon requires a minimum of three squads (each having three teams) to accomplish the mission. The other MP platoons are normally employed forward in sector, providing AS and force protection for the division main and tactical (TAC) CPs and for other HVAs in the vicinity of the maneuver brigade rear boundary. As with the AOE heavy division, the Division XXI heavy division requires augmentation with at least one corps MP company for CS missions in the DRA and can expect two companies in a mature theater (based on METT-TC) as the main effort. In addition, the division may request an MP battalion headquarters to command and control CS MP operations in the rear or to conduct TCF operations.

7-37. While the division MP company is capable of performing all five MP functions, it must be augmented by the corps MP to fully perform all of these functions throughout a mature division's battle space. The PM prioritizes MP functions (based on the supported commander's needs), conducts a mission analysis, and requests needed corps assets through the division G3. Corps MP companies, L&O detachments, and CID assets are needed for sustainment operations. When corps MP companies and L&O detachments are attached to the division, they are placed under OPCON of the PM. The PM will then assign an AO and missions based on METT-TC.

7-38. The PM supports the division's distribution-based logistics concept by conducting aggressive R&S operations that secure the LOC. Since the main effort is to secure the LOC, traditional MMS missions (such as convoy security and escorts) are performed by exception. Through the allocation of increased MP to conduct LOC R&S operations aggressively, the division is able to detect and destroy threat forces before they interrupt operations in rear-area or sustainment operations.

7-39. Digitization gives the division the situational awareness that allows it to cover much larger distances and to move much faster. The systems that provide this digitization are generally unprotected HVAs on the battlefield, which require aggressive AS and force-protection operations.

7-40. The EPW operations are much more fluid, sometimes requiring the division GS MP to move forward and assume brigade collection points, thus allowing DS platoons to move forward with their brigades. Similarly, corps MP may be required to move forward and assume the division's central collection points as the division's rear boundary is pushed forward to shorten the LOC.

7-41. As with all AOE divisions, until corps L&O and CID detachments augment the division, L&O operations are on an exception basis. The MP collect intelligence during every facet of mission execution. The PM develops police intelligence in the division main CP through coordination with the SOC, the G2, the Assistant Chief of Staff, G5 (Civil Affairs) (G5), and the Assistant Chief of Staff, G6 (Communications) (G6).

Chapter 8

Separate Brigades and Initial/Interim Brigade Combat Teams

The MP support separate brigades and IBCTs designed to provide the brigade commander with MP assets that can serve as a force multiplier and that can be employed as an economy of force. The brigade MP platoon is as lethal, flexible, and capable as any other platoon in the MP Corps.

USAMPS

The Army uses separate brigades to inject a small but powerful force where it is needed. It must be able to fight and win while operating on its own for a sustained period of time. It must be able to defend itself on a 360-degree front in war or MOOTW. While there are some variations of separate brigades, it is the heavy separate brigade that is most commonly found within the force.

OVERVIEW

8-1. The Army's IBCT is a full-spectrum, wheeled combat force. It is employed in all operational environments against all projected future threats. However, it is designed and optimized primarily for employment in small-scale contingency operations in complex and urban terrain, confronting low-end and midrange threats that may employ both conventional and asymmetric capabilities. The IBCT deploys very rapidly, executes early entry, and conducts effective combat operations immediately on arrival to prevent, contain, stabilize, or resolve a conflict through shaping and decisive operations. The IBCT participates in war (with augmentation) as a subordinate maneuver component within a division or a corps and in a variety of possible roles. The IBCT also participates (with appropriate augmentation) in stability and support operations as an initial-entry force or as a guarantor to provide security for stability forces by means of its extensive capabilities.

SUPPORT TO SEPARATE BRIGADES

8-2. MP support to separate brigades is provided by a four-squad MP platoon organic to the brigade HHC. A separate PM cell within the brigade HHC serves as the C^2 element for the platoon (see Figure 8-1, page 8-2). Support to the platoon and the PM section for maintenance, supply, mess, and communications is provided by the brigade HHC. Since the platoon and PM section have no organic support, the MP leadership must perform close coordination for this support. However, the MP platoon must compete with other brigade HHC assets for priority of repair for weapons, vehicles, and communications equipment.

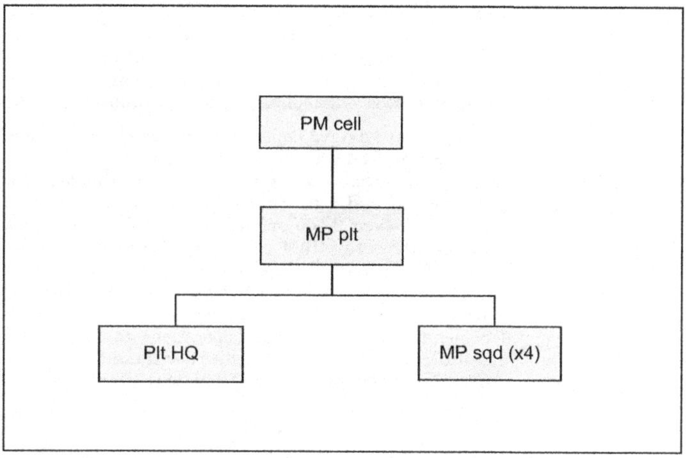

Figure 8-1. MP Support to a Separate Brigade

8-3. The C^2 of an MP unit supporting separate brigades also extends down from the tactical commander. The separate brigade PM has OPCON of brigade MP assets the same way the division PM has OPCON of division MP assets. This includes any MP assets that may have been provided from the corps. The PM advises the commander of a separate brigade on matters pertaining to MP operations. The platoon leader directs the execution of his platoon's mission based on the priorities set forth by the PM and the supported commander. Since the separate brigade's organic MP platoon is more robust than an MP platoon supporting a division maneuver brigade, METT-TC will determine the requirement for augmentation. However, corps L&O and CID augmentation is required.

8-4. The separate brigade's MP platoon is capable of performing all five MP functions. However, its resources are quite limited. Although the MP squads are employed according to METT-TC, the platoon supporting the separate brigade may have—

- One squad operating in the EPW collection point.
- One squad providing a mobile security screen and providing AS around the brigade's main CP.
- Two squads conducting MMS and AS throughout the brigade's rear area.

SUPPORT TO THE INITIAL/INTERIM BRIGADE COMBAT TEAM

8-5. The IBCT is a divisional brigade. It is designed to optimize its organizational effectiveness and to balance the traditional domains of lethality, mobility, and survivability with the domains required for responsiveness, deployability, sustainability, and a reduced in-theater footprint. Its two core qualities are its high mobility (strategic, operational, and tactical) and its ability to achieve decisive action through a dismounted infantry assault. The major fighting components are its motorized infantry battalions. The IBCT has a unique reconnaissance, surveillance, and target acquisition (RSTA) squadron to enhance situational understanding.

8-6. To achieve a rapid deployment threshold, the brigade's design capitalizes on the widespread use of common vehicular platforms—particularly a highly mobile, medium-weight, combat/CS platform coupled with the minimization of the personnel and logistical footprint on theater. Encompassing a personnel strength of about 3,500 and preconfigured in ready-to-fight combined-arms packages, the entire IBCT can deploy within 96 hours of "first aircraft wheels up" and begin operations immediately upon arrival at the APOD. In essence, the APOD is the TAA. The IBCT cannot conduct forced-entry operations, but it provides the JFC with an improved capability to arrive immediately behind forced-entry forces and begin operations to shape the battle space and to execute decisive action to expedite decisions. Once committed, the IBCT can sustain operations for up to 180 days without relief.

8-7. The IBCT's organization is expandable either through augmentation or scalability (according to METT-TC) in any given contingency. The IBCT is scalable in terms of its ability to accept like-type additional forces to expand core tasks and functions already resident within the IBCT (for example, adding additional infantry or RSTA organizations). The IBCT is also capable of accepting augmentation consisting of units or elements that execute tasks or functions not resident within the IBCT (for example, adding armor, air-defense, MP, or aviation assets). In both cases, added units execute their normal mission-essential task list (METL) tasks and, therefore, will not require extensive training to deploy and operate with the IBCT. In both cases, the IBCT includes the command, control, and communications (C^3) capabilities necessary to permit the rapid integration of additional enabling capabilities, particularly for operations outside the scope of small-scale contingency operations (such as stability operations, support operations, and war).

8-8. The IBCT organization excludes other unit-based capabilities often provided in a division slice. However, for each missing capability, the brigade headquarters includes staff cells capable of conducting rudimentary planning and analyses to ensure that all functional-area considerations are incorporated in route planning and preparation for operations. The first MP elements in support of the IBCT are at the brigade headquarters level.

8-9. The MP planning cell is composed of a two-person (a major [MAJ] and a sergeant first class [SFC]) planning cell located within the HHC's maneuver-support cell and under the direct staff supervision of the brigade S3. The role of the MP planning cell is significantly different from that of a traditional division PM or a separate brigade PM. The main difference is in the IBCT's lack of organic or habitual MP assets. However, the absence of organic MP assets makes the job of the MP planning cell that much more critical. The MP planning cell must—

- Understand the organization, capabilities, and limitations of the IBCT.

- Conduct effective liaison with higher headquarters PM elements.
- Become an effective planner and anticipator of MP requirements.
- Prepare MP staff estimates and employment recommendations.
- Plan for MP deployment via air, sea, rail, or land.
- Task-organize MP units effectively and efficiently.
- Assume C^2 of incoming MP forces or, if operating under a division or corps, relinquish C^2 to the division/corps PM or battalion commander (if appropriate for effective employment of MP forces).

8-10. Depending on METT-TC, the brigade may be augmented by MP elements ranging from a platoon to a battalion. In any case, the MP planning cell then becomes a staff planner and coordinator of functional matters pertaining to—

- Force protection, physical security, and vulnerability countermeasures.
- Operations for collecting, processing, and evacuating (possibly) EPWs, CIs, and detainees.
- US military prisoner operations.
- Operations for processing and controlling dislocated civilians and refugees.
- Customs and counterdrug operations.
- Criminal investigations and CID-/MP-investigator support.
- PIO and establishing links with HN police agencies and other international/interagency law-enforcement agencies.
- MMS, AS, and L&O operations.
- Coordination of MP or CID division/echelons above division (EAD) augmentation forces, MWD teams (explosives, narcotics, and patrol), L&O detachments, and MP I/R units through the division, corps, and major Army command (MACOM) PMs.
- Training guidance to other US forces required to perform MP functions in the absence of MP forces.

8-11. Since the IBCT is a divisional brigade, the division PM and the IBCT MP planning cell play an important role in developing an optimum MP force package to support the brigade commander's concept of operations. Despite the brigade's early-entry timelines, the MP planning cell must consider and plan for MP augmentation forces as early as possible to free valuable combat resources. Small-scale contingency operations that result in numerous EPWs, civilian detainees, and refugees will hamper momentum and freedom of maneuver.

8-12. Once the initial brigade receives MP augmentation (see Figure 8-2), the MP priority of effort during the offense will focus on ensuring that routes remain unencumbered and secure for movement of ground combat, CS, and CSS forces. The MP's priority of effort during the defense will focus on conducting AS and counterreconnaissance along the LOC, C^2 centers, and CSS bases. The MP may conduct response-force operations or become part of the TCF.

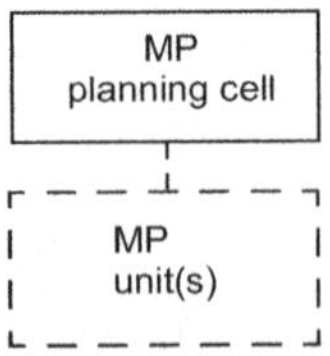

Figure 8-2. MP Support to the IBCT

8-13. Stability and support operations present some unique challenges. The IBCT may be deployed to a geographical area that is politically unstable, that lacks civil control, or that is in complete turmoil. The MP planning cell plans for and coordinates MP support according to available resources and the supported commander's needs. In this scenario (and based on METT-TC), an MP battalion TF could be required to deal with the challenges of—

- A significant number of refugees or dislocated civilians.
- AS or force-protection issues.
- MMS operations.
- Black-market and criminal investigations.
- Restoration of order.
- Civil unrest.
- Intervention of private and nongovernmental organizations.

Chapter 9

The United States Army Criminal Investigation Command

The history of the USACIDC goes back to World War I, when General John J. Pershing organized the CID in France. Today, the USACIDC is the Army's sole agent responsible for investigating felony crimes on and off the battlefield. The USACIDC provides investigative support to commanders at all echelons.

USACIDC

The USACIDC investigates offenses committed against US forces or property, or those committed by military personnel or civilians serving with US forces or where there is a military interest. The USACIDC agents investigate violations of international agreements on land warfare. They conduct special investigations at the direction of the USACIDC's commanding general (CG) or a higher authority. In general, the USACIDC's missions include—

- Investigating and deterring serious crimes.
- Conducting sensitive/serious investigations.
- Collecting, analyzing, processing, and disseminating criminal intelligence (CRIMINTEL).
- Conducting protective-service operations for designated personnel.
- Providing forensic-laboratory support.
- Maintaining Army criminal records.
- Enhancing the commander's crime-prevention and force-protection programs.
- Performing LOGSEC operations.

OVERVIEW

9-1. The USACIDC's operations help the commander maintain discipline and order by preventing or investigating crimes that reduce a unit's ability to fight. During the investigation of serious crimes, the USACIDC concentrates its efforts on investigating serious crimes such as wrongful deaths, controlled-substance offenses, theft, fraud, sex crimes, and assaults. The USACIDC also conducts sensitive and special investigations involving matters pertaining to senior Army officials and those associated with classified programs.

9-2. The USACIDC agents collect, analyze, process, and disseminate criminal intelligence/information relating to crime within or directed toward the Army. Specific information relating to modus operandi, crime techniques, investigative leads, gang violence, and terrorism is shared with the appropriate intelligence and law-enforcement agencies. Conversely, USACIDC agents solicit and receive crime-related information from the MP and from local, national, and foreign law-enforcement and intelligence agencies. Special agents identify and evaluate crime-conducive conditions and indicators of potential attacks against Army property, facilities, or personnel. They then provide reports to the appropriate commander.

COMMAND AND CONTROL

9-3. The USACIDC is a centralized (stovepipe) MACOM whose special agents in the field report through the USACIDC's chain of command (detachment to battalion to group) to the CG, who in turn reports directly to the Army Chief of Staff and the Secretary of the Army (see Figure 9-1).

9-4. As an independent MACOM, the USACIDC's tactical units are not under the C^2 of supported organizations. The USACIDC's elements may be temporarily attached to a supported commander when required to accomplish a nonstandard mission. Reassignment, promotion, accreditation, and disciplinary actions are retained by the USACIDC. Attachments will be coordinated with the appropriate USACIDC headquarters planning the specific mission and approved by the CG, USACIDC. Additionally, although there is no formal staff relationship, USACIDC commanders advise their supported commanders on criminal-investigation matters. This enhances the quality, reliability of information, support, and trust between USACIDC elements and their supported commanders.

WARTIME SUPPORT

9-5. The USACIDC supports each echelon of command from the division to the ASCC. The theater USACIDC structure is comprised of a C^2 headquarters

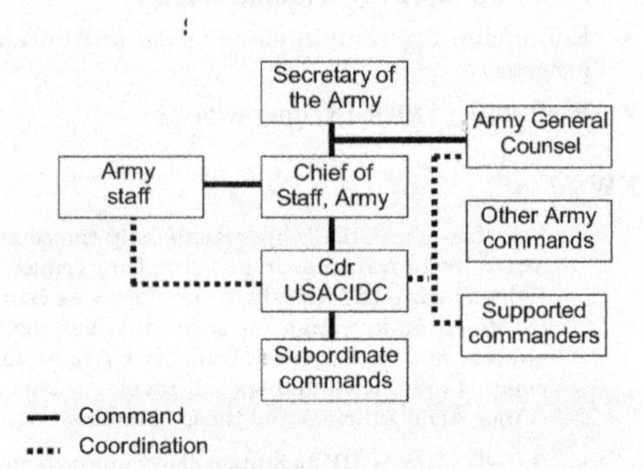

Figure 9-1. USACIDC Chain of Command

and mobile, modular, and tailorable investigative detachments. The USACIDC supports combatant commanders with the following functions:

- **LOGSEC.** Tracking and protecting materials and equipment from the manufacturer to the soldier on the battlefield.
- **CRIMINTEL.** Collecting, consolidating, analyzing, and disseminating intelligence associated with criminal and terrorists activities targeted at Army interests.
- **Criminal investigations.** Investigating war crimes and, in some cases, crimes against coalition forces and HN personnel.
- **Protective-service operations.** As in peacetime, protecting key personnel anywhere on the battlefield.

9-6. The USACIDC's LOGSEC function protects the Army's supply pipeline against criminal activities from the manufacturer, through logistics channels, all the way to the frontline soldier. It involves preventing, detecting, and investigating criminal and terrorist activities such as supply diversion, destruction, and sabotage or product substitution. The USACIDC's LOGSEC operations assess LOGSEC, identify weaknesses, and provide a prioritization of threats so that commanders can implement preventive measures to reduce the vulnerability of the logistics pipeline. Whenever possible, the USACIDC will initiate actions to recover logistical losses and return them to Army control.

> The Army let a $22 million contract for the manufacture of its mobile kitchen trailer (MKT)-90 mobile field kitchens in 1990. These kitchens were distributed to Army units worldwide, including those units serving in Saudi Arabia during Operation Desert Shield/Desert Storm. After distribution of the MKT was well underway, a US government quality-assurance representative received information from the manufacturer's employees indicating that the kitchens were not built to contract specifications. This information led to an investigation by the USACIDC's major fraud procurement unit (MFPU). The investigation revealed that inferior material and manufacturing processes were substituted, which allowed the contractor to realize an additional $228,000 in profit. The US Army Environmental Hygiene Agency determined that the MKT-90 had serious sanitation defects and posed a safety hazard to troops using and being fed from these kitchens. The problem with the MKT was identified and expeditiously resolved to ensure the safety of soldiers in the field. The primary subject was indicted by a grand jury and subsequently pled guilty to one count of major fraud. He was found guilty, sentenced to 10 months confinement, and ordered to pay restitution.

9-7. The USACIDC's CRIMINTEL collection efforts focus on the identification and prevention of terrorist and nonterrorist crimes against US and allied military personnel, facilities, and other interests. Upon collecting this information, agents recommend countermeasures to combat subversive activities through coordination with MI, the PM, rear-area operations officers, and HN military and civil intelligence agencies. Additionally, when directed, the USACIDC becomes the lead US military investigative agency at theater level. When this happens, it is tasked with leading the prevention-of-terrorism effort from all services, not just the Army component.

> The MI analysts, the MP, and US Army criminal investigators deployed in Kosovo formed a combined all-source information center. This center provided threat assessments to units assigned to TF Falcon. Fifty-five percent of these assessments referenced some form of criminal activity. These intelligence products provided commanders with reliable information concerning potential incidents or criminal activity. With nearly 7,000 US soldiers deployed within the region, these assessments were absolutely crucial for the force-protection planning effort.
>
> The fusion cell within the information center developed intelligence products from national, theater, and operational sources. Due to the significant threat posed by criminal elements in the region, two CID military agents and two CID civilian analysts were attached to the fusion cell to facilitate the police-intelligence function. The CID personnel, in cooperation with MP soldiers, played a key role by linking CRIMINTEL to specific groups and events. The CRIMINTEL collection effort was specifically targeted on weapons, drugs, organized crime, and the identification of smuggling routes. The identification of smuggling routes resulted in a significant increase in the number of weapons being confiscated. The timely transfer of CRIMINTEL products to tactical units enabled a rapid response to serious confrontations, increased confiscation of arms and ammunition, and improved stability in TF Falcon's AO.

9-8. During war and MOOTW, the criminal-investigation effort includes—

- War crimes and, when directed, crimes against coalition forces and HN personnel.
- Major felony crimes committed by US personnel or EPWs/CIs and, when directed, crimes committed by foreign nationals in which there is an Army interest.
- Criminal acts by indigenous personnel, factions, and ad hoc groups.
- Special/sensitive investigations, hostage negotiations, polygraph operations, and force-protection operations.
- CRIMINTEL operations.

9-9. The USACIDC detachments provide enhanced coverage and protection for designated key and essential leaders during war and MOOTW. The protection requirement for senior JTFs and Army commanders may be significantly greater during MOOTW than during peacetime or war as the propensity of asymmetrical threats (such as criminal and terrorist groups) operating in the AO also increases.

ORGANIZATION

9-10. The USACIDC is organized with an array of multifunctional units. As with many other Army organizations, its force is in a transition process. During peacetime, the USACIDC groups provide brigade-level area support for worldwide Army operations. The group provides C^2, staff planning, and administrative support for USACIDC elements assigned to an AOR or deployed to a contingency operation. During war, a group provides the same capabilities in support of an ASCC and provides C^2 for all USACIDC units in the theater. Regardless of the operational environment, the group ensures the connectivity between all USACIDC units. It establishes and maintains links with supported units and interagency, joint, multinational, allied, and HN authorities on matters pertaining to Army and USACIDC operations. The group headquarters has a command section, a detachment headquarters, an S1, an S2, an S3, an S4, an SJA, and a communications section. It provides C^2 for up to six subordinate battalions. Major differences in the AOE and Force XXI groups are in an increase of support and mission personnel and the movement of all polygraph support down to battalion level.

9-11. During peacetime, the USACIDC battalions provide area support for worldwide Army operations. They provide C^2, staff planning, forensic-science support, CRIMINTEL, polygraph support, technical guidance and supervision of USACIDC operations, and administrative support for USACIDC elements assigned to an AOR or deployed to a contingency operation. During war, the USACIDC battalion provides the same capabilities in support of each corps and TSC. Under AOE designs, one battalion supports each corps and TSC. Under the Force XXI design, up to two battalions support each TSC. Like the groups, each USACIDC battalion ensures connectivity and establishes and maintains links with other units/agencies within its AOR. The battalion headquarters consists of a command section, a detachment headquarters, an S1, an S2, an S3, an S4, and a communications section. It provides C^2 to five USACIDC detachments on a sustained basis. Its span of control can accommodate up to seven detachments for a short period of time. Major differences in the new battalion are the addition of a CSM, an S6 officer, a detachment commander and support personnel, and CRIMINTEL and forensic personnel.

9-12. The USACIDC investigative team is the smallest operational investigative element. Each team consists of two special agents (a warrant officer and a noncommissioned officer [NCO]). METT-TC requirements dictate that these teams have the capability to operate independently from the detachment headquarters. The investigative teams are the building block for both AOE and Force XXI USACIDC detachments. Both designs provide levels of flexibility to task-organize without splitting units apart. The detachment (with its complement of teams) performs the full range of criminal-investigative functions in all operational environments, but it has limited CRIMINTEL and LOGSEC management capabilities and no forensic or polygraph capability. Each detachment includes a headquarters section and up to eight investigative teams. METT-TC requirements dictate that these teams have the capability to operate independently from the detachment headquarters. The major differences between the AOE and Force XXI detachments' design begin with replacing the heavy and light division support elements (DSEs), the corps-area support element (CASE), the TSC-area support element, (TASE), and the port-area support element (PASE) designs with a modular design consisting of a standardized headquarters element and two special-agent sections. Each section consists of four investigative teams. Personnel administration, supply, and the bulk of administrative support found in the AOE organizations have been consolidated at battalion.

Additionally, the Force XXI detachment is commanded by a special-agent warrant officer.

9-13. Both the AOE and the Force XXI units retain the Army's standard dependencies upon other units for support requirements (religious, health services, finance, photographic processing, and so forth). The new USACIDC TOEs have enhanced A/L support capabilities. However, the new units still retain the AOE support requirement for food services and emergency Class III/V resupply from CS MP brigades, battalions, and companies.

THE FIELD INVESTIGATIVE UNIT

9-14. Sensitive investigations are normally conducted by the field investigative unit (FIU). The FIU is a one-of-a-kind organization within the DOD that enables the Secretary of the Army to conduct sensitive investigations requiring access to special information or programs that are highly classified. The FIU works closely with the Army IG, the Judge Advocate General of the Army, and the Army General Counsel to support commanders in the special-operations and intelligence communities and in the area of classified acquisition programs. The FIU may also be assigned investigations involving senior Army personnel or those of special interest to the Army leadership.

THE COMPUTER-CRIME INVESTIGATION UNIT

9-15. The computer-crime investigative unit (CCIU) deals with intrusions involving classified networks or multijurisdictional offenses. The CCIU works closely with MI and federal law-enforcement agencies to coordinate military actions, nonmilitary-affiliated offenders, and foreign-intelligence services. In addition, it provides technical assistance to CID elements that are conducting computer-related investigations.

THE PROTECTIVE-SERVICE UNIT

9-16. The protective-service unit (PSU) provides worldwide protective services to designated personnel to protect them from assassination, kidnapping, injury, or embarrassment. The PSU plans, coordinates, and executes executive protection for—

- The Secretary of Defense.
- The Deputy Secretary of Defense.
- The Chairman of the Joint Chiefs of Staff.
- The Vice Chairman of the Joint Chiefs of Staff.
- The Secretary of the Army.
- The Army Chief of Staff.
- The Army Vice Chief of Staff.
- Others (as directed).

9-17. During war and MOOTW, the USACIDC routinely provides protective-service support to senior Army ground commanders and, in some cases, to JTF commanders. The USACIDC agents also provide training to the MP details providing protective services to senior field commanders.

The President of the US hosted the North Atlantic Treaty Organization's (NATO's) 50th summit anniversary celebration in Washington, DC during the period of 22 to 26 April 1999. The President invited 28 heads of state or prime ministers, 48 foreign ministers, 88 ministers and chiefs of defense, and the chairman of the military committee to the summit. During the conference, the principal guests resided in 23 hotels in Washington, DC. The responsibility for providing protection for the ministers, chiefs of defense, and the chairman for the military committee was given to the USACIDC's PSU. In preparation for this mission, additional training on protective-service operations and evasive driving techniques was given to over 200 reserve agents. These individual mobilization augmentees, troop program units, and other USACIDC agents were critical to the successful execution of this mission.

US ARMY CRIMINAL-INVESTIGATIONS LABORATORY

9-18. Forensic-laboratory support is provided by the US Army Criminal-Investigations Laboratory (USACIL) that is currently located at Fort Gillem, Georgia. The USACIL is a nationally accredited forensic laboratory that provides all three services with dedicated, highly responsive, deployable, state-of-the-art scientific and forensic investigative support. The six forensic divisions of the USACIL include the following:

- Chemistry (serology, drugs, trace, and deoxyribonucleic-acid [DNA] testing).
- Questioned documents (alterations and comparisons of handwriting, typing, and printing).
- Firearms and tool marks (weapons and obliterated numbers).
- Photography (evidence, charts and photos for court).
- Latent prints.
- Computer forensics.

9-19. The USACIL collects, receives, and examines evidence and prepares reports of findings. It also provides expert-witness testimony at court-martials.

US CRIME RECORDS CENTER

9-20. The US Crime Records Center (CRC) maintains the Army's criminal records. It maintains more than two million MP reports and USACIDC reports of investigations. These records are retained for at least 40 years before being destroyed. During their retention, the records are actively maintained and frequently researched.

During the course of an investigative search, civilian police authorities seized 19 hand grenades. A routine search of stolen property and criminal index files failed to identify the grenades as having been stolen from a nearby military installation. However, a search of the CRC files identified a USACIDC investigation relating to the theft of a case of grenades that occurred 20 years earlier. Further investigation revealed that the lot number of the recovered grenades was identical to that of the stolen grenades. The investigative leads resulting from the CRC file search contributed to the resolution of both cases.

9-21. The CRC conducts more than 10,000 criminal-history name checks each month to identify victims and perpetrators of criminal offenses. The checks are requested not only by USACIDC agents, but also by other military and civilian law-enforcement officials. The CRC is also the Army's agent for Freedom of Information Act (FOIA) and Privacy Act requests relating to criminal investigations and MP reports. The CRC also manages the Army's polygraph programs and their support to Army installations around the world.

EMPLOYMENT

9-22. Whenever possible, USACIDC units are colocated with the MP. This provides unified MP support across the battlefield, and it facilitates logistical support (primarily mess) provided by MP units. However, the USACIDC maintains its vertical C^2. Based on METT-TC, the MP group (CID) will colocate with the supported ASCC and the MP battalions (CID) supporting the corps and EAC will colocate with the MP brigade (CS). When feasible, MP detachments (CID) will colocate with MP battalions/companies and division MP companies. Figure 9-2 illustrates a typical battlefield array for AOE organizations, and Figure 9-3 illustrates a Force XXI array.

9-23. Since USACIDC detachments are austere organizations with little organic A/L capability, they rely on the battalion for unit-level maintenance, supply, and personnel support. However, detachments must often operate independently when performing criminal-investigation functions throughout the AOR. During these times of long-distance separation, detachments must seek A/L support from the supported unit.

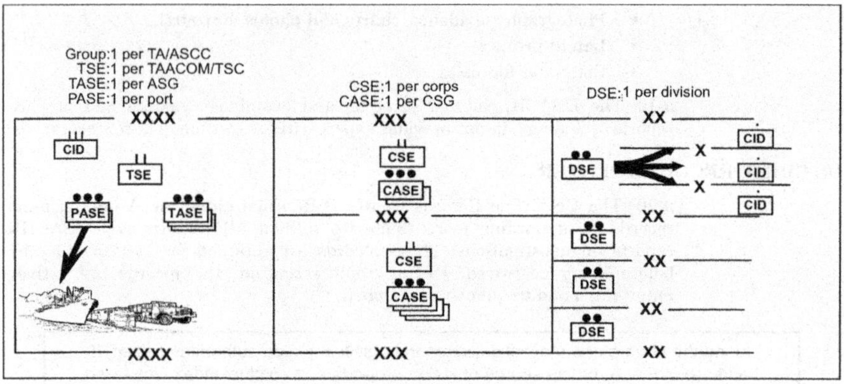

Figure 9-2. AOE Battlefield Array

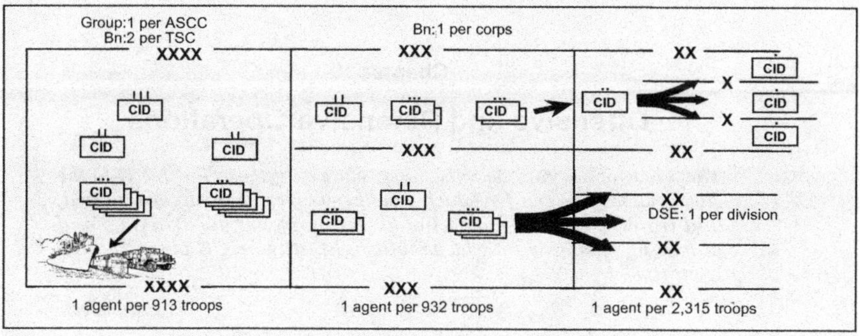

Figure 9-3. Force XXI Battlefield Array

Chapter 10

Offensive and Defensive Operations

The whole of military activity must relate directly or indirectly to the engagement. The end for which a soldier is recruited, clothed, armed, and trained; the whole object of his sleeping, eating, drinking, and marching is simply that he should fight at the right place and the right time.

Clausewitz

Army doctrine considers the full range of operations from war to MOOTW. An operation is a military action or the carrying out of a military mission needed to gain the objectives of any battle or campaign. Offensive and defensive operations normally dominate military operations in war and in some smaller-scale contingencies. An offensive operation is aimed at destroying or defeating the enemy and imposing our will on him for a decisive victory. Defensive operations defeat an enemy attack, buy time, economize forces, or develop conditions favorable for a counteroffensive that regains the initiative and defeats the enemy.

OVERVIEW

10-1. The offense is the decisive form of battle. Offensive campaigns and major operations are designed to achieve operational and strategic objectives quickly and decisively at the least cost. Offensive operations combine the terrain and the force to achieve the objective. The main purpose of the offense is to defeat, destroy, or neutralize the enemy force. Offensive operations are characterized by a rapid shift in the main effort to take advantage of opportunities by momentum and simultaneous destruction of enemy defenses.

10-2. The immediate purpose of defensive operations is to defeat an attack. Military forces defend only until they can gain sufficient strength to attack. Commanders choose to defend when they need to buy time, to hold a piece of terrain, to facilitate other operations, or to erode enemy resources at a rapid rate while reinforcing friendly operations. The ARFOR conduct defensive operations as part of major operations and campaigns and in combination with offensive operations (see FM 3-0).

SUPPORT TO OFFENSIVE OPERATIONS

10-3. MP support to offensive operations varies according to the type of operation being conducted. The MP functions support each echelon commander based on METT-TC. In the offense, MP priorities are often placed on MMS, I/R, and AS operations. Regardless of the offensive effort, MP leaders supporting an offensive operation must—

- Understand the commander's intent and his concept of operations.
- Anticipate that selective elements of the offensive force may need to pause, defend, resupply, or reconstitute while other forces attack.
- Anticipate changes in the operational tempo and prepare to weight the MP effort to meet the maneuver commander's intent.
- Provide MP support within the context of linear and nonlinear operations. They must understand how both types of operations affect the security of rear areas and LOC, the protection of C4I, the means of sustainment, and the land force's mobility.
- Support the maneuver commander's intelligence efforts by conducting the PIO function. The MP must observe, identify, and report information.
- Assist maneuver forces to mass. The MP protect C^2 nodes such as the CP that orchestrates the mass efforts and resources that sustain the offensive move. The MP must deny incursions by enemy reconnaissance forces into the AO. They also maintain surveillance, provide early warning, and impede and harass the enemy with supporting and organic fires.
- Speed the commitment of follow-on forces and reserves. They must also conduct MMS operations to assist the commander in a smooth and quick shift of forces to support the main effort.
- Anticipate the pursuit and exploitation by positioning MP forces that can support continuous operations. The MP support as far forward as possible while protecting resources needed to seize the opportunity. The EPWs/CIs are collected as far forward as possible. The MP coordinate the treatment of dislocated civilians and refugees with the HN or foreign forces.
- Tailor the MP support based on the unit being supported (corps, divisions, and brigades).
- Consider the location and composition of response forces and the TCF and the strength of rear-area threats.
- Integrate a force into the offensive plan that is capable of conducting aggressive combat operations against designated targets.

> During Operation Just Cause, an MP platoon received the mission to disable an armed Panamanian Defense Force (PDF) patrol boat docked near a key objective. After receiving the mission, the platoon leader maneuvered his force into position and attacked after receiving enemy fire. The platoon successfully disabled the patrol boat and captured the objective while suffering minimum casualties.

10-4. During offensive operations, MP support provided by teams, squads, and platoons provide the commander with an agile, flexible, versatile, and capable force ready to contribute to the overall mission success.

SUPPORT TO DEFENSIVE OPERATIONS

10-5. The MP support defensive operations consistent with the commander's intent and priorities. They support a defending force by assisting subordinate elements in the execution of different, yet complementary actions. For example, the MP may support the maneuver of an attacking unit by conducting MMS operations while simultaneously providing AS for a defending unit in a base cluster. The MP support reflects the maneuver commander's focus—destroying an attacking force, retaining or denying key terrain, moving away from an enemy force, or a combination of the aforementioned. However, despite the intent of the defensive operation being conducted, MP leaders must—

- Conduct MMS operations to help a force maneuver and mass. The MP must anticipate the shift from the defense to the offense and assist the forward movement of reserves.
- Conduct AS operations to deny information to enemy reconnaissance elements seeking out the exact location of the defending force. The MP are positioned where they can control key terrain or improve the defensive capability of bases and base clusters. The MP conduct aggressive R&S to locate and deny enemy access to critical logistical facilities.
- Protect sustainment resources while supporting the lateral, forward, and rearward movement of forces.
- Consider the type and size of the AO, the LOC security, the rear threat, and the number of EPWs/CIs and dislocated civilians to determine how they may affect the movement of forces.

> During Operation Desert Shield, EAC, corps, and division MP units conducted a myriad of missions to support both the XVIII Corps's (Airborne) and the VII Corps's movement to their TAA. The MP's main focus was on MMS and AS throughout the AO to ensure freedom of movement and to protect US and coalition forces against terrorists and other threats in the rear area.

10-6. The success of MP operations in the defense depends on the MP leader's situational awareness, the commander's intent, and the precise employment of MP resources when and where they are needed. In the defense, the MP perform their five functions to assist the defending commander by providing a lethal mobile force that allows him to concentrate his efforts and resources in fighting the next battle.

Chapter 11

Stability and Support Operations

ARFOR conduct stability operations in a complex, dynamic, and often asymmetric environment. Stability operations are usually nonlinear and noncontiguous. They are often time and manpower intensive. The purpose of support operations is to meet the immediate needs of designated groups for a limited time until civil authorities can accomplish these tasks without Army assistance.

FM 3-0

The Army's mission is to prepare for war and, when deterrence fails, to achieve a quick, decisive victory. The DOD projects that many of its future missions will be stability and support operations. However, these operations are not new to the Army. The Army has participated in stability and support operations in support of national interests throughout history. The Army has protected its citizens, assisted nations abroad, and served America in a variety of other missions. The pace, frequency, and variety of stability and support operations have quickened in the last three decades. As demonstrated in Operations Provide Comfort, Restore Democracy, and JTF Los Angeles, these types of operations present a challenge to all services.

OVERVIEW

11-1. Stability operations are conducted outside of the US and its territories to promote and protect US national interests by influencing political, civil, and military environments and by disrupting specific illegal activities. Stability operations may include both developmental and coercive actions. Developmental actions enhance a HN government's willingness and ability to care for its people. Coercive actions apply carefully prescribed limited force or the threat of forces to change the AO's environment.

11-2. Support operations provide essential supplies and services to relieve suffering and to help civil authorities prepare or respond to crises. In most cases, the ARFOR's efforts are focused on overcoming conditions created by man-made or natural disasters. The ultimate goal of support operations is to meet the immediate needs of designated groups (for a limited time) until civil authorities can accomplish these tasks without Army assistance.

11-3. Stability and support operations are distinctly different from offensive and defensive operations and must be analyzed differently. For instance, the application of the rules of engagement (ROE) will be unique to the operation. The restrictions on combat operations and the use of force must be understood and obeyed at all levels. MP leaders must clearly take the initiative and ensure that the applicable guidance is implemented and followed by all subordinates.

11-4. Stability and support operations have more diverse political considerations than are usually encountered in war. There are also complexities that must be addressed in joint, combined, and interagency stability and support operations. The Army's challenge is to execute these operations successfully in a joint or combined arena without degrading its capability to wage war successfully.

11-5. If the US is to fight and win future wars, it must develop, train, and employ its force structure in a manner that ensures success in every operation across the spectrum of military operations. Under these auspices, MP forces can provide commanders with especially valuable assets for stability and support operations. The MP have the unique capability of serving as a combat, CS, and CSS force. The domestic and international acceptability of the MP operational image frequently makes the MP the most appropriate force for stability and support operations. Additionally, the MP are flexible and capable of rapidly transitioning from one end of the spectrum to the other if the environment changes.

FORCE SUITABILITY

11-6. The MP's capabilities and their operational and supporting tasks are as integral to stability and support operations as they are to offensive and defensive operations. MP branch-/core-related skills are highly compatible with the capabilities required for stability and support operations. The MP provide a highly capable, politically acceptable force that is suitable for a variety of missions. They possess robust moving, shooting, and communicating capabilities; and they project an assist, protect, and defend image that is particularly important when tailoring a force that requires significant capabilities but a low political profile. The MP provide a flexible, but limited, economy-of-force organization. Mission requirements will always exceed available MP resources, especially in stability and support operations. For this reason, MP assets must be prioritized to deliver the greatest mission capability. The specific operations that MP units perform at a given time are determined by the echelon commander's needs, the mission requirements, and the availability of MP resources. As discussed in previous chapters, the echelon commander, through the command's PM, sets the priorities for MP operations. As in offensive and defensive operations, the PM knows that while the MP force can perform all MP functions, they may not be robust enough to perform all assigned MP missions all of the time. The PM must continuously make trade-offs between the numbers and kinds of MP missions that can be supported.

11-7. The MP's capabilities are further enhanced by their training, mind-set, and experience in dealing with people in highly stressful and confusing situations. Although capable of conducting combat operations when needed, the MP are highly practiced in de-escalation and in employing the minimum-essential force to contain potentially violent situations. This mind-set serves as the framework for MP law-enforcement training and is especially applicable in MP support for stability and support operations. Furthermore, this mind-set is exercised and reinforced daily in peacetime law-enforcement operations. This is a significant capability. MP soldiers learn and receive constant reinforcement training in controlling a situation on the spot before it escalates and cannot be controlled. The MP have a unique understanding of human nature and are adept at handling emotion-laden situations.

> During operations in Kosovo, as both Serbs and Albanians attempted to create unfavorable conditions and responses by US forces, the MP kept their cool under fire in the streets of Gnjilane and Kamenica. The MP's reaction led to a quick understanding by all sides that the US Army MP were fair and just and understood the tense situations.

11-8. The MP stress four essential training competencies—skill proficiency, human dimension and attitude, camaraderie and teamwork, and leadership. These competencies are the foundation for all MP operational performances. They equip the MP to exercise discretion in dealing with others, to protect and assist those in need, and to accomplish the varying demands of MP missions in stability and support operations and, when needed, in war.

STABILITY AND SUPPORT

11-9. Stability and support operations are performed in support of US authorities or in support of international bodies and foreign governments when so required by US authorities. The MP have long provided the essential support for stability and support operations; however, the frequency of these operations has increased significantly in recent years. Consequently, the MP have been deployed in greater numbers with greater frequency. MP units have a substantial history of successful, simultaneous, and continuous deployments in both a lead and a support role.

SUPPORT TO DOMESTIC CIVIL AUTHORITY

11-10. The purpose of support to domestic civil authorities is to—

- Render humanitarian assistance.
- Provide disaster relief.
- Restore order.
- Combat terrorism.
- Support counterdrug operations.

11-11. Military operations supporting domestic civil authorities are governed by the provisions of the Stafford Act, the Posse Comitatus Act, and other laws and regulations. Peacetime experience in civil-military affairs equips the MP to oversee, assist, and train other units and agencies for law enforcement, sensitive security, and operations to control dislocated civilians (within the legal limits of a particular stability or support operation). For this reason, the MP are often needed after other forces have redeployed, and they stay until government order is restored. Since active-component military missions are constrained by the Posse Comitatus Act, nonfederalized national guard (NG) units are often employed to conduct these operations (particularly counterdrug operations). The NG units can operate under gubernatorial authority within state jurisdiction without Posse Comitatus restrictions. The NG MP units are the only NG units trained, skilled, and experienced in providing law enforcement. They offer civilian authorities a unique, specialized capability not available from other NG units.

SUPPORT TO HUMANITARIAN ASSISTANCE AND DISASTER RELIEF

11-12. Humanitarian assistance and disaster-relief operations provide emergency assistance to victims of natural and man-made disasters and other endemic conditions. Natural disasters include earthquakes, floods, and hurricanes. Man-made disasters include riots and civil conflicts. Humanitarian assistance and disaster relief conducted by US forces alleviate urgent nonmilitary needs of a populace until the appropriate civilian agencies can provide the needed services without military support. These types of operations include, but are not limited to—

- Conducting searches and rescues.
- Evacuating, sheltering, sustaining, and protecting dislocated civilians.
- Making food, medical care and treatment, and other civilian-aid programs available.
- Reducing opportunities for criminal activity and restoring L&O.
- Maintaining other operations needed to ensure the well being of the affected population.
- Providing relief operations for man-made disasters. These operations are characterized by civil conflict, and they entail all of the requirements for other disasters but focus on the restoration of L&O.
- Providing force protection and security.

11-13. MP support can be an important asset in all disaster-relief operations. The MP are trained and equipped for decentralized operations. MP teams operate in highly mobile vehicles equipped with radios, which are invaluable in disaster-relief operations. Each MP platoon, company, and battalion headquarters is equipped with high-frequency radios. These capabilities allow the MP (even when widely dispersed) to maintain centralized communications; to provide L&O support; and to assist in search, rescue, and evacuation efforts. Because of their decentralized operations and communications equipment, the MP can also play a vital role in notification, ADC, and access and egress control. They can deploy with disaster-assessment teams to assist in advising and developing a disaster-assistance plan. The MP can also deploy to an affected area to—

- Assist in restoring order.
- Search for, rescue, and evacuate victims.
- Provide personnel and vehicular circulation controls.
- Provide ADC.
- Conduct detention operations.
- Provide physical and area security to protect life and property immediately following a disaster.
- Protect fire and emergency-services personnel.
- Gather, record, and report information.
- Provide force protection and security.

11-14. MP (I/R) battalions can support civil-affairs units by establishing I/R facilities to shelter, sustain, protect, and account for disaster-relief victims. These units can also train and assist government agencies (US and foreign) or police in the operation of facilities for dislocated civilians and detainees.

11-15. The MP possess capabilities critical to relief in man-made disasters characterized by civil conflict. The MP have the essential training and skills needed for relief operations that focus on restoring civil order. Restoration of civil order demands—

- **Objectivity and neutrality.** The objective of civil-disturbance control is to restore L&O, not to impose punishment or suppress peaceful dissent. Government forces quelling civil disorders must be emotionally objective and politically neutral. MP law-enforcement training stresses objectivity and neutrality, an important attribute for forces employed in stability or support operations.
- **Minimal use of force.** The application of more than minimum force may be detrimental to mission success. The MP are particularly suited for these situations, having been trained on the technique for the use and measured application of minimum forces to accomplish mission goals.
- **Public acceptance.** Government forces must avoid actions that will alienate the populace. Any action offensive to community sensitivities must be avoided. The MP receive training in crowd-control operations, interpersonal communications skills, and minimum use-of-force techniques—skills that are critical in gaining public acceptance.
- **Positive image.** In many disturbance operations, demonstrators and government forces compete for the approval and sympathy of the general population. The MP's assist-and-protect image provides the government with a distinct advantage in such situations.
- **Threat awareness.** Government forces must pay attention to and try to learn the organization, intentions, and activities of demonstrators. Equal attention must be paid to how the community will respond to the demonstrator's actions and the government's response. The MP interact daily with the populace, placing them in a position to develop criminal information and to determine the collective attitude of the populace.
- **Effective coordination.** Disturbance operations often involve many agencies from various levels of government. Thorough, timely, and coordinated preparation, planning, training, and execution are essential for mission success. MP civil-disturbance training (including planning and coordination measures) and their communications assets enhance coordination, particularly during mission execution.
- **Operational unity.** Multiple control forces representing different agencies with varying jurisdictional authority can pose C^2 problems. When unity of command cannot be achieved, operational unity must be achieved by colocating operational centers, integrating communications, and delineating organizational responsibilities. The MP are well trained in police and military operations, possessing a sound understanding of both civilian police and military measures, legal constraints, and effective coordination.

11-16. The MP operational tasks supporting man-made disasters vary according to the nature of the disaster and the operational environment. For example, in CONUS where the commitment of military forces to civil disturbances is considered a last resort and military involvement is limited by law, nonfederalized NG MP units have long been employed by civil authorities for their unique capabilities. MP forces can—

- Isolate and contain an area with barricades, roadblocks, and perimeter patrols. The MP can also assist civil-affairs units by enforcing pass and identification systems and public-utility controls.
- Provide mobile patrols to enhance the security of high-priority targets (such as buildings, utilities, and services that are critical to the community's economic and physical well being).
- Monitor, disperse, contain, or limit crowd movement through—
  - Observation points and patrols.
  - The communication of interest and intent to the crowd.
  - The channelization or diversion of the crowd.
  - Gaining the cooperation of the dissident leadership.
  - The proclamation of the illegal nature of the crowd's actions.
  - Show of force.
  - The use of crowd-control formations.
  - The apprehension of crowd members and leaders.
  - The application of minimum-force measures.
- Establish area control to prevent looting, to protect businesses and other likely targets, and to prevent arson. The MP accomplish this mission with saturation patrolling, including vehicle and foot patrols. With augmentation, the MP can perform air and water patrols. They also enforce populace control measures (such as ordinances to prevent gathering, permits to gather, restrictions on circulation, restrictions on the interference with government and public functions and personnel, restrictions on possessing weapons, and other measures instituted by civil authorities).
- Neutralize special threats (such as snipers and bombings) that are highly dangerous to both government forces and the community. Snipers pose a particularly grave danger in disaster-relief operations. The MP can take immediate protective actions, secure an area, isolate the threat and, when authorized, use SRTs to assist civilian or HN authorities in apprehending or neutralizing the sniper. The MP are trained to perform the necessary actions when an explosive device is discovered. In such an event, they can secure and evacuate an area, organize search teams, and isolate the site.

11-17. The MP can also support in rescue and recovery operations, which may include rescuing US or foreign nationals. Such operations are also employed to locate, identify, and recover sensitive materials deemed critical to national security. The operations may be conducted in benign or hostile environments. The MP provide OPSEC, and they process and account for civilians affected by rescue and recovery operations. The MWD teams can also help search for lost or hidden personnel. In stability and support operations, the MP can perform their police functions for rescue and recovery operations while keeping the perceived military signature low.

SUPPORT TO COUNTERDRUG OPERATIONS

11-18. Counterdrug operations are measures taken to disrupt, interdict, and destroy illicit drug activities. In a counterdrug role, the DOD may offer certain forms of support to HN counterdrug personnel, to US civilian law-enforcement agencies operating in a HN, to Department of State (DOS) counterdrug personnel, and to the Bureau of International Narcotics Matters. Without direction from the National Command Authority acting under constitutional or statutory authority, US forces engaged in counterdrug activities may not engage in direct law-enforcement activities (such as apprehension, search, seizure, and other similar activities). As discussed earlier, the Posse Comitatus Act authorizes specific DOD assistance in counterdrug activities, but US military forces usually may not participate directly in arrests, searches, seizures, and similar activities. Conversely, as discussed earlier, USC Title 32 authorizes state governors to use NG MP units with full authority to engage directly in law-enforcement activities. When authorized, US advisors may accompany counterdrug forces on operations.

11-19. MP operational contributions can include the use of force, drug identification, crime-scene protection, customs operations, surveillance of criminal activity, serious-incident reporting, and other related activities. Employing the MP for these operations ensures force suitability and avoids a mismatch in terms of unit capability and mission effectiveness.

11-20. Border-screening operations consist of three separate but related measures—mobile patrolling, the use of listening/observation posts, and airfield surveillance. These measures are part of the MP's AS and MMS functions used in tactical rear-area operations. The MP's operational effectiveness may be further increased through augmentation by a civilian law-enforcement officer who would be responsible for search, seizure, and arrest actions. Active-component units are permitted to assist in marijuana eradication in support of domestic law-enforcement agencies (DLEAs). Marijuana-eradication operations are an application of the MP AS, mobility, and L&O functions, thus providing realistic training while supporting an operation.

11-21. As discussed in previous chapters, the MP are well suited to augment customs operations. They routinely perform customs inspections for US forces in Korea and Germany. The NG MP regularly conduct vehicle, cargo, and container searches under the supervision of US Customs agents and at US ports of entry.

11-22. The USACIDC agents can provide support to counterdrug operations. The CID units identify the sources of illegal-drug distribution systems. When directed, the MP and the CID complement interagency counterdrug efforts to stop the production, flow, and distribution of drugs and to provide planning, training, equipment, and facilities to support domestic counterdrug operations. The CID also supports CRIMINTEL programs, internal security, and site surveys at borders where Army units will deploy. In addition to the aforementioned support, the DOD General Counsel has ruled that MWD teams can be loaned to DLEAs to assist in detecting smuggled contraband.

SUPPORT IN COMBATING TERRORISM

11-23. The Department of Justice (DOJ) is the lead federal agency for dealing with acts of terrorism within the US, its territories, and its possessions. Within the DOJ, the Federal Bureau of Investigation (FBI) has the lead. The FBI can train the police of friendly nations in antiterrorism and counterterrorism operations. Usually, US military forces act in a technical-advisory but not a law-enforcement capacity in combating acts of terrorism. It is often difficult to distinguish the acts of politically motivated terrorists from violent acts of criminals or individuals in society at large. In fact, all terrorist acts are criminal; however, they also may be part of an insurgency.

11-24. The MP are key players in the defensive planning, implementation, and control of combating-terrorism operations, whether in garrison or deployed to war or MOOTW. In whatever countries US forces operate, MP units conduct antiterrorism operations daily as part of their L&O and AS functions. The MP also respond to terrorist incidents. They help secure the area and search for evidence, and they help the lead agency. The MP provide advice and assistance in developing antiterrorism and counterterrorism programs for deployed US forces and DOS agencies and for police, paramilitary, and military forces of assisted nations (when so authorized). MP training, liaison, and joint patrols help strengthen HN-police organizations, improving their effectiveness and efficiency in combating terrorism. The MP also assist HN police in performing PRC operations.

11-25. The training of police (domestic and foreign) requires DA approval, and restrictions exist on some forms of training. Legal advice and coordination are a must before training can occur. The MP and the CID provide personnel and equipment for planning, advising, equipping, and training agencies to prevent, deter, and respond to terrorism. The training focuses on deterring terrorist acts such as bombings, kidnappings, assassinations, taking hostages, and hijackings. Related assistance may include training and equipment for explosive detection, management of hostage situations, physical security, protection of key personnel, and defensive/evasive driving methods. For counterterrorism operations on US military installations, the MP provide highly trained, specially equipped SRTs for situations requiring the selective use of force. Additionally, CID agents are highly trained in hostage negotiation, which increases the commander's options before resorting to force.

11-26. The MP and the CID assist in counterterrorism information-gathering activities through liaison and joint patrols with HN police. These activities are part of the PIO function. The MP have extensive contacts with civilian police, who in turn have contacts with the populace. These contacts can help surface criminal information concerning terrorist plans, recruitment, and support structures sooner than military sources. The CID can provide commanders with CRIMINTEL on a daily basis. At the same time, MP antiterrorism support includes measures to reduce the vulnerability of people and property. Such measures include—

- Personal awareness.
- Personal-protection operations and techniques.
- Crime prevention.
- Physical-security programs.

11-27. The MP and CID crime-prevention programs employ proactive measures to protect people and their property and US and HN property. Crime prevention is a continuous process of planning, implementing measures, evaluating effectiveness, modifying measures, and developing further information. The primary crime-prevention tool is the crime-prevention survey. It is used to examine the physical features of various critical sites. The MP physical-security programs help identify, reduce, eliminate, or neutralize conditions favorable to criminal, terrorist, and insurgent activities. The key measures used to accomplish this are physical-security inspections and surveys.

SUPPORT TO NONCOMBATANT EVACUATION OPERATIONS

11-28. Noncombatant evacuation operations (NEOs) relocate civilians who are endangered in a foreign nation. The NEOs are classified as benign (unopposed) or hostile (opposed). MP roles in support to NEOs differ accordingly. In a benign environment, NEOs are usually conducted with the help and full cooperation of the affected nation, with little or no opposition to evacuation. The use of force is limited to self-defense (as outlined by the ROE) and protection of the evacuees. In a hostile environment, NEOs are opposed by forces with specific intentions of preventing or destroying the operation. In this scenario, the MP will follow the ROE as established by higher headquarters.

11-29. MP operations in NEOs closely parallel EPW/CI evacuation operations. The objective of both is to safely collect, secure, protect, transport, process, and account for people, whether they are EPWs or noncombatants. However, due to the dissimilarities of these groups, the techniques employed differ.

11-30. The MP can secure selected facilities, reception and transfer points, or other locations required to collect, process, and evacuate people. They can provide route reconnaissance and mobile security teams to escort HRP, DOS personnel, evacuees, and sensitive material. If needed, MP units assist civil-affairs personnel with the movement and control of dislocated civilians. The MP may support civil affairs by setting up and operating evacuation sites, holding areas for foreign nationals denied evacuation, and reception and processing stations. This support can be provided by MP CS units for a short duration and by MP I/R units for a prolonged duration. Additionally, the MP will establish and maintain liaison with HN police to ensure a mutual understanding of jurisdiction, to coordinate efforts, and to provide criminal and tactical intelligence. In hostile environments requiring the employment of combat forces, the MP would be employed in their traditional CS role.

SUPPORT TO SECURITY-ASSISTANCE OPERATIONS

11-31. Security-assistance operations are one of the main tools of US foreign policy. A critical concern in these operations is the HN's ability to plan and manage its own defense resources. In security-assistance operations, the US provides defense materiel, military training, and defense-related services through grants, loans, credit, or cash sales. When authorized, the MP support security assistance with military and police training through the International Military Education and Training Program (IMETP). This program provides instruction and training to foreign military and civilian personnel on a grant/aid basis. Many foreign civil law-enforcement agencies request training in the following areas:

- Basic police procedures (such as patrolling and crime-scene protection).
- Physical security.
- Corrections.
- Civil-disturbance operations.
- Customs operations.
- Traffic control.
- Use of force.

11-32. The MP possess the expertise and experience to advise, train, and assist a HN's military and police forces. The MP's participation in the IMETP includes—

- Formal and informal instruction of foreign students.
- Technical education and tactical training with applicable training aids.
- Assistance to foreign police and military elements by MP mobile training teams (MTTs).
- Providing MP units or small, modular, functionally specific teams to support security-assistance needs. These teams can train HN police in both field and institutional facilities.
- Support to nation-assistance operations.

11-33. Nation-assistance operations support a nation's effort to promote development, ideally by using its national resources. This is achieved through education and the transfer of essential skills to the HN. The goals of such operations are to—

- Promote long-term regional stability.
- Develop sound and responsive democratic institutions.
- Develop supportive infrastructures.
- Promote an environment that allows for orderly political change and economic progress.

11-34. The MP support nation-assistance operations with training and joint MP and HN-police operations. The MP know and understand police operations, military operations, and C^2 procedures. Consequently, the MP are ideal for training the law-enforcement personnel in both field and institutional settings. The MP's nation-assistance goals are to—

- Enhance L&O in democratic societies.
- Improve efficiency and effectiveness.
- Promote the proper usage and maintenance of equipment.
- Establish a sound training base for police.

- Standardize procedures that enhance combined police operations.
- Promote friendship and goodwill toward the US.
- Preclude the need for US military interventions to counter acts of violence.

11-35. Combined police operations reinforce training and provide HN police with mobility, security, and communications to operate more effectively while conserving critical personnel resources. The MP provide support through training and the following:

- Law-enforcement, security, and criminal-information support to the HN police force (to include L&O operations and administration in a democratic society).
- Patrol and desk operations.
- Circulation-control operations.
- PRC and civil-disturbance operations.
- MWDs.
- Physical-security operations.
- Personnel-security operations.
- Area- and route-security operations.
- Counterdrug operations.
- Antiterrorism operations.
- Mass immigration operations.
- Customs and border operations.
- Confinement operations.
- Crime prevention.
- Community and police relations.
- C^2 of police operations.

11-36. When conducting nation-assistance operations, as well as other operations, the MP will often be tasked to support civil-affairs operations. Close coordination with the G5/S5 and liaison with the appropriate HN representatives have added emphasis to these types of operations. The scope of civil-affairs operations is influenced by the economic, social, psychological, and political background of the country and its people. Civil-affairs personnel will coordinate the MP's effort to restore stability, contribute to national development, and promote support for the host government. Civil-affairs personnel assist in preventing civilian interference with military operations and coordinate all other military and civil affairs (such as community relations, PRC, civil defense, and community- and area-security programs).

11-37. The MP can conduct ADC measures to assist the HN in the aftermath of a natural or man-made disaster. They help rescue and evacuate the injured for medical care, and they issue food, water, and essential supplies (See FM 41-10).

SUPPORT TO PEACEKEEPING OPERATIONS

11-38. Peacekeeping operations (PKOs) are military operations conducted with the consent of the belligerents in a conflict. The PKOs are intended to maintain negotiated truces and to facilitate diplomatic solutions. The US may participate in a PKO under the auspices of an international organization, in cooperation with other countries, or unilaterally. A PKO usually occurs in an ambiguous situation in which there is extreme tension and the possibility of violence. This violence may range from terrorism and sabotage to minor military conflicts involving known and unknown belligerents. MP forces may serve as an element of a UN peacekeeping force or as part of a multinational observer group. The overall operational control of a peacekeeping force is exercised by a multinationally staffed military peacekeeping command. However, the commander of each assigned national force retains command integrity of his unit.

11-39. The basic MP force structure and appropriate augmentation are situation dependent. The size and composition of the MP forces are determined by diplomatic negotiations. Personnel spaces are allocated based on the US contingents' respective missions. The principle of consent affects the composition and function of the peacekeeping force. Consent applies to the degree of interest the disputing countries have when contributing their forces to participate in the peacekeeping effort.

11-40. Each PKO is unique but is generally characterized by constraints in the use of force. In a PKO, force is normally limited to self-defense, the defense of others, and the protection of national security. Therefore, the soldiers in a PKO must possess the skills required for warfare, but they must also be trained so that they are equally able to act as intermediaries. In this aspect, the MP are a distinct asset. They are experienced in a case-by-case approach to the use of force. They are experienced in using discretion and in the de-escalation of potential violence. The MP training in EPW operations are especially useful in handling EPWs if the peacekeeping force supervises the exchange of prisoners of war.

11-41. The peacekeeping force employs a combination of the following techniques to control the potential for violence inherent to PKOs:

- Observation is the peacekeeping force's primary responsibility and basic requirement. It provides timely and accurate reports of any suspicious situation, incident, or occurrence.
- Information gathering is a continuous requirement. Peacekeepers must be constantly alert to what takes place around them and to any changes or inconsistencies in the behavior, attitude, and activities of military and civilian personnel.
- Surveillance and supervision operations help oversee the implementation of agreements. The operations require restraint, tact, and patience.
- Complaint investigations must be thorough and objective, and they must result in a fair assessment.
- Negotiation and mediation skills are often required of peacekeepers. If peacekeepers can resolve minor problems at the lowest level, they can prevent major issues from rising, which may prevent increased tensions or the resumption of fighting.
- Patrolling (mounted and dismounted) among the population publicizes presence. It also serves as a means of gathering information, informally enhancing supervision, and investigating complaints.

11-42. The above techniques are compatible with MP battlefield functions. Operationally, the MP conduct route and area R&S, operate mounted or dismounted patrols, operate listening/observation posts, provide humanitarian assistance, investigate possible terrorist and criminal acts and cease-fire and sanction violations, and gather information. The MP provide a flexible, wide-ranging means of information dissemination. Additionally, they provide crucial support for force protection. The MP and the CID employ active and passive measures to deter and defeat threats directed at service members, their families, DOD civilians, facilities, equipment, and very important persons (VIPs). The MP use emerging technologies in force protection to conduct security, detection, and identification operations and to warn of intruders.

11-43. As part of a PKO, the MP conduct crowd- and riot-control operations, including the extraction of mob leaders. The MP are an ideal force for controlling antagonistic masses engaged in rioting, looting, and demonstrating. The MP understand how to make the transition from the lower end of the use-of-force spectrum to the use of lethal force, if so required or directed. Finally, the MP deploying with peacekeeping forces can provide early-on force protection, including headquarters and initial aerial port/seaport of debarkation (APOD/SPOD) security. The MP's mobility, firepower, and communications provide critical reconnaissance, information-collection, and response-force capabilities and acceptability not found in combat forces.

11-44. When conducting crowd- and riot-control operations in a PKO or peace-enforcement operation (PEO), nonlethal weapons (NLWs) are an additional means of employing force for the purpose of limiting the probability of death or serious injury to noncombatants or belligerents. However, the use of lethal force must always remain an inherent right of individuals in instances when they, their fellow soldiers, or personnel in their charge are threatened with death or serious bodily harm. The NLWs add flexibility to the control of disturbances by providing an environment where military forces can permissively engage threatening targets with limited risk of noncombatant casualties and collateral damage (see FM 90-40).

11-45. The use of lethal force (employed under the standing ROE) will never be denied. At no time will forces be deployed without the ability to defend themselves against a lethal threat nor will they forego normal training, arming, and equipping for combat. Nonlethal options are a complement to, not a replacement for lethal force. They seek to expand a proactive response across the range of military operations.

11-46. The decision to use NLWs against an adversary during a confrontation should be delegated to the lowest possible level, preferably to the platoon or squad. However, this requires that all personnel (not just leaders) have a clear understanding of the ROE and the commander's intent.

11-47. Commanders and public-affairs officers must be prepared to address media questions and concerns regarding the use and role of NLWs. They must be prepared to address the role of NLWs, and they must make it clear that the presence of NLWs in no way indicates abandoning the option to employ deadly force in appropriate circumstances.

11-48. The PEOs are in the high end of the PKO's spectrum. These operations are just short of combat and require a different operational approach. MP operations to support PEOs correspond to MP battlefield functions supporting combat forces. The PEOs often pose special operational considerations for dealing with dislocated civilians. The MP provide PRC operations and I/R, when required.

ORGANIZATIONS AND CAPABILITIES

11-49. MP units are well suited for stability and support operations given their unique capabilities and low force signature. The MP brigade, battalion, company, and I/R units may be tailored for stability and support operations based on their unique METT-TC.

MP BRIGADE TF

11-50. In essence, the organization and capabilities of the MP organization in stability and support operations is the same as those for other operations. However, its augmentation is tailored based on METT-TC unique to stability and support operations and the commander's needs. The MP brigade has been the centerpiece of successful MP deployment during previous stability and support operations. The MP brigade provides comprehensive planning, C^2, and A/L support to assigned and attached subordinate units. The brigade also has the command and staff experience to coordinate and supervise HN security and law-enforcement activities.

11-51. The MP brigade headquarters provides senior MP leadership and the communications capabilities required to direct a variety of functional elements—CSS, PSYOP, public affairs, and combat-arms augmentation—in support of joint or unilateral TF operations. At authorized level of organization (ALO) 1, the MP brigade headquarters has an SJA section to coordinate the legal and jurisdictional issues inherent to stability and support operations. The brigade headquarters also has an S5 section and a liaison capability to coordinate support and assistance to HN agencies, friendly forces, other US forces, DOS personnel, and nongovernmental or private volunteer organizations.

11-52. The MP brigade TF (MPBTF), with its robust capabilities and low force signature, is well suited for the unique mission requirements and sensitivities of stability and support operations. The MPBTF can be employed as part of a combat TF in high-threat, opposed-entry operations or as the lead TF for a low-threat, unopposed-entry operation. The following are examples of previous MPBTFs and the operations in which they participated:

- **Operation Provide Comfort.** The 18th MP Brigade deployed to northern Iraq to provide humanitarian assistance and protection to the Kurds.
- **JTF Los Angeles.** The 49th MP Brigade (NG) provided civil-disturbance assistance to Los Angeles in the aftermath of the Rodney King trial.
- **Operation Hawkeye.** The 16th MP Brigade and units of the 89th MP Brigade provided security, protection, and humanitarian

assistance to the population of Saint Croix in the aftermath of Hurricane Hugo.

- **JTF Andrew.** The 16th MP Brigade provided humanitarian assistance in the aftermath of Hurricane Andrew.
- **Operation Sea Signal.** The 89th MP Brigade deployed to Guantanamo Bay, Cuba, to provide humanitarian assistance to Haitian migrants.
- **Operation Restore Democracy.** The 16th MP Brigade deployed to Haiti to provide humanitarian assistance.

11-53. Early deployments of the MPBTF headquarters or a mission-tailored portion of the headquarters positions a senior, effective MP C^2 headquarters to respond to all MP requirements. It can assist in the coordination with the media, governmental and nongovernmental agencies, and other forces, and it can coordinate civil-military aspects of the operation. The MP brigade headquarters can assimilate and employ additional forces without degrading operations. The augmentation of brigade headquarters is driven by METT-TC. With the brigade headquarters in place, subordinate or supporting units can be significantly expanded without the loss of C^2. In essence, little or no loss in operational momentum occurs when the C^2 element deploys first.

11-54. The MP LRP section of the MP brigade headquarters colocates and works with the higher headquarters' G3/Operations Directorate (J3) plans element. This allows the MP LRP to incorporate MP operations effectively into higher-level planning, and it enables an efficient transition to field operations. Also, because in peacetime this section functions as a PM cell, it provides the PM with efficient and effective technical support during stability and support operations. As the operations approach termination or enter sustainability and fewer forces are required, the brigade headquarters can redeploy, leaving a battalion TF in charge of MP forces.

MP BATTALION TF

11-55. The MP battalion (CS) provides command, control, and support for assigned MP companies (CS). MP battalions (CS) are best employed as part of a TF—either as part of a combat TF (high threat, opposed) or a support TF (low threat, unopposed). With sufficient augmentation, the battalion can provide the logistical, administrative, personnel, and operational support needed to lead a diverse TF. During stability and support operations, an MP battalion TF may be placed under the OPCON of a brigade-size TF (MP or non-MP) or a division TF. If placed under the OPCON of a division TF, both the division PM and the MP battalion commander will synchronize their efforts to maximize the employment of available divisional and nondivisional MP assets in support of the division commander's concept of operations.

11-56. As with the MPBTF, the MP battalion's contributions to stability and support operations are not a new concept. The following are examples of previous battalion TFs and the operations in which they were involved:

- **Cuban Refugee Movement.** The 519th, 716th, 720th, and 759th MP Battalions deployed for humanitarian-assistance operations.
- **Operations Restore Hope and Continued Hope.** The 720th MP Battalion provided humanitarian assistance in Somalia.

- **Operation Sea Signal.** The 720th MP Battalion deployed to Guantanamo Bay, Cuba, to conduct humanitarian-assistance operations for Haitian migrants. Later, the 716th and 759th MP Battalions also deployed to Guantanamo Bay to perform like functions for the Cubans.

MP COMPANY IN STABILITY AND SUPPORT OPERATIONS

11-57. The MP company is the MP Corps's primary mission performer. It executes a wide variety of missions. In these operations, MP companies—

- Provide mobility support for both vehicles and personnel.
- Provide AS, including security for critical personnel, sites, cargoes, and railways.
- Secure, safeguard, shelter, and control detainees and dislocated civilians.
- Restore order in civil disturbances and conduct PRC operations.
- Carry out limited L&O operations.
- Provide liaison, coordination, and training for all aspects of law enforcement to other agencies.
- Conduct police-intelligence collections and disseminate information to the populace, friendly forces, and other HN agencies.
- Support force protection and security.

11-58. The MP company possesses robust mobility, lethality, and communications. Each of the MP companies' three-man teams are equipped with either a high-mobility, multipurpose, wheeled vehicle (HMMWV), an up-armored HMMWV, or an armored security vehicle (ASV); an AN/VRC 91 radio; a Light-Vehicle Obscuration Smoke System (LVOSS); an MK-19 grenade machine gun; a squad automatic weapon (SAW); two M-16 or M-4 rifles; an M203 grenade launcher; and a 9-millimeter pistol as a side arm for each team member. In addition to conducting combat operations, the MP company commander has the capability to conduct limited L&O operations (desk and investigative operations). With a tremendous operational reach, the MP company will displace its platoons as far as possible to accomplish all of the MP functions.

MP I/R UNITS IN STABILITY AND SUPPORT OPERATIONS

11-59. The MP I/R units are specifically equipped and trained to support, safeguard, account for, and provide proper and humane treatment for EPWs/CIs, US military prisoners, or dislocated civilians. When conducting dislocated-civilian operations, the MP I/R units support and interact with civil-affairs and PSYOP units who are responsible for monitoring and handling dislocated civilians. Should more than one category of personnel have to be supported, the MP I/R units will establish separate compounds for each to preclude contact with other populations. When operation of the facility is going to be more than a short-term endeavor, the intent is to involve (as appropriate) NGOs, IHOs, HN authorities, and dislocated civilians in the administration of the facility. The tasks performed by MP I/R units for stability and support operations are basically the same as those for war, but the operations and procedures are modified according to the operational environment and the categories of people to be housed in the facility (see FM 19-40).

11-60. Criminal detainees (who are detained against their will) are subject to controls and procedures that closely parallel those used for EPWs/CIs. The primary emphasis is on guarding and preventing escape until their disposition is determined by recognized civilian judicial/law-enforcement authorities. When the operation is conducted outside of the US and its territories, criminal detainees are treated according to established legal procedures.

11-61. Dislocated civilians (who are being assisted rather than detained) are provided aid, shelter, and protection. The emphasis is on protecting them from harm by natural forces or hostile personnel. A special category of personnel arises when the MP I/R unit is required to house personnel that are dislocated civilians but who must be detained against their will. Such is the case of mass migrants who flee their countries and find themselves under US custody while formal proceeding guidance is being developed. In this case, MP I/R units must be sensitive to the situation and attempt to strike a balance between security, shelter, protection, and detention procedures.

11-62. Military detainees (who are detained against their will) are subject to controls and procedures that closely parallel those used for EPWs/CIs. In stability and support operations, US forces detain opposing-force military personnel. Without a formal declaration of war, opposing military personnel who are captured cannot be categorized as EPWs. However, they are afforded many of the EPW's rights and privileges under the Geneva convention. Such a situation occurred during Operation Just Cause when PDF personnel were detained by US forces.

11-63. In an operational environment in which hostile groups are engaged against one another, an I/R facility may be set up to protect one group from another. In this case, the facility's purpose is to shelter, sustain, account for, and protect its dislocated civilians from the hostile group. The MP I/R units and other supporting units concentrate on providing AS to protect the facility from direct fire. Other MP or combat forces provide protections beyond the direct-fire zone. Accountability for the dislocated civilians is coordinated with the SJA and civil affairs. They focus on maintaining a record of the people in the facility and their physical conditions. In operations in which no hostile groups are engaged (such as natural disasters), the I/R facility may be set up to provide shelter, sustain, and account for personnel. There may not be a need for external security personnel.

11-64. The MP I/R C^2 structure for stability and support operations is based on METT-TC. The nature and complexity of the mission, the number and types of detainees, and the operational duration should be considered. At one end of the operations spectrum, an MP brigade (I/R) may be required; while at the other end, an MP battalion (I/R) may be assigned to an MP brigade (CS).

Chapter 12

Force Protection

There is no more important responsibility than force protection. It is an inherent command responsibility and must be fully integrated into every unit's mission. A commander must continually review his unit's force-protection posture and crosswalk it with current and changing policy and threat levels. Force protection demands the personal involvement of commanders to ensure the best possible security consistent with the threat to personnel and mission-essential resources.

AR 525-13

Force protection consists of those actions that prevent or mitigate hostile actions against DOD personnel (to include family members), resources, facilities, and critical information. It coordinates and synchronizes offensive and defensive measures to enable the joint force to perform while degrading opportunities for the enemy. It includes air, space, and missile defense; NBC defense; antiterrorism; defensive-information operations; and security to operational forces and means.

OVERVIEW

12-1. The MP contribute significantly to the Army's overall force-protection effort. They are directly responsible for executing certain security-related measures (law enforcement and physical security) and for providing support to others (HRP security and antiterrorism), all of which contribute to force protection. Commanders use the Army's risk-management methodology and established security programs and capabilities within US forces to help prioritize and implement force-protection measures.

SUPPORT TO FORCE PROTECTION

12-2. MP support to force protection is accomplished under the leadership of the respective command's PM. It is accomplished mainly through an active role in physical security, HRP security, law enforcement, and antiterrorism.

PHYSICAL SECURITY

12-3. Physical security is an integral part of security of forces and means. Physical-security measures deter, detect, and defend against threats from terrorists, criminals, and unconventional forces. These measures include—

- Fencing and perimeter standoff space.
- Lighting and sensors.

- Vehicle barriers.
- Blast protection.
- Intrusion-detection systems and electronic surveillance.
- Procedural measures such as security checks, training and awareness, property accountability/inventory requirements, physical-security inspections of mission-essential or vulnerable areas (MEVAs), and physical-security surveys of installations.

12-4. These physical-security measures are not exclusively peacetime measures. They are designed to protect US personnel and equipment regardless of the mission or the geographical location.

12-5. The MP support the commander's security of forces and means programs by—

- Controlling or monitoring installation, base, or base-cluster access or entrance points.
- Monitoring intrusion-detection systems and providing a response force.
- Conducting physical-security inspections.
- Conducting perimeter security or site surveillance.
- Recommending the placement of walls, berms, gates, or barriers around designated MEVAs, high-value areas, or perimeters.
- Supporting the commander's risk-analysis effort.
- Conducting roving patrols, checkpoints, or roadblocks.
- Performing other physical-security measures as required by the commander.

HRP SECURITY

12-6. The HRP security program provides additional security to designated individuals (and their families) whom by virtue of their rank, assignment, symbolic value, vulnerabilities, or location or a specific threat are at a greater risk than the general population. This security program is applicable across the full spectrum of military operations.

12-7. The MP support the commander's HRP security measures by—

- Conducting protective-service details for HRP (normally conducted by USACIDC special agents).
- Responding to duress alarms and intrusion-detection systems.
- Conducting a PSVA (performed by the USACIDC).
- Employing MWDs.
- Coordinating with local, state, and federal agencies as well as joint, multinational, and HN law-enforcement agencies.
- Performing other HRP security measures as required by the commander.

LAW ENFORCEMENT

12-8. The Army's law-enforcement program supports the security of forces and means through the prevention, detection, response, and investigation of crimes. The MP, the USACIDC, and DOD police play a crucial role in the law-enforcement program. They provide a visible deterrent against a broad portion of the threat spectrum, investigate crimes, and provide an initial response force to threat incidents. As with the previous programs, the law-enforcement program is applicable anywhere across the spectrum of military operations.

12-9. The MP support law-enforcement operations by—

- Providing liaison teams with local, state, and federal agencies; HN police; and joint and multinational agencies.
- Employing SRTs and hostage-negotiation teams (normally provided by the USACIDC).
- Providing traffic enforcement, MSR regulation enforcement, and other route-control measures.
- Employing MWDs.
- Conducting MP investigations (criminal and noncriminal).
- Conducting patrolling, AS, and surveillance measures.
- Implementing applicable threat-condition (THREATCON) measures.
- Conducting and implementing other law-enforcement measures as required by the commander.

ANTITERRORISM

12-10. Terrorism is characterized as the unlawful use of violence or the threat of violence to coerce or intimidate a government or a society. Protection against a terrorist threat requires both an offensive counterterrorism capability and a defensive antiterrorism program. The primary focus of antiterrorism is on training and awareness and thorough planning. Antiterrorism is not a stand-alone program. It leverages other force-protection-related programs (such as physical security and law enforcement) to provide much of the physical defense against terrorism.

12-11. The MP support the commander's antiterrorist program by—

- Collecting CRIMINTEL (within the limits of AR 190-45).
- Responding to terrorist threats and incidents.
- Performing aggressive patrols and R&S operations.
- Detecting suspicious activities.
- Enforcing joint and cooperative agreements with the FBI, the USACIDC, MI, and other pertinent agencies (home or abroad).
- Conducting information collection and dissemination.
- Protecting critical assets, facilities, and personnel.
- Performing other antiterrorism measures as required by the commander.

Appendix A

Metric Conversion Chart

This appendix complies with current Army directives which state that the metric system will be incorporated into all new publications. Table A-1 is a conversion chart.

Table A-1. Metric Conversion Chart

Metric to English			English to Metric		
Multiply	**By**	**To Obtain**	**Multiply**	**By**	**To Obtain**
Length					
Centimeters	0.0394	Inches	Inches	2.54	Centimeters
Meters	3.28	Feet	Feet	0.0305	Meters
Meters	1.094	Yards	Yards	0.9144	Meters
Kilometers	0.621	Miles (stat)	Miles (stat)	1.5609	Kilometers
Kilometers	0.540	Miles (naut)	Miles (naut)	1.853	Kilometers
Millimeters	0.039	Inches	Inches	25.40	Millimeters
Area					
Square centimeters	0.1550	Square inches	Square inches	6.45	Square centimeters
Square meters	10.76	Square feet	Square feet	0.0929	Square meters
Square meters	1.196	Square yards	Square yards	0.836	Square meters
Volume					
Cubic centimeters	0.610	Cubic inches	Cubic inches	16.39	Cubic centimeters
Cubic meters	35.3	Cubic feet	Cubic feet	0.0283	Cubic meters
Cubic meters	1.308	Cubic yards	Cubic yards	0.765	Cubic meters
Milliliters	0.0338	US liq ounces	US liq ounces	29.6	Milliliters
Liters	1.057	US liq quarts	US liq quarts	0.946	Liters
Liters	0.264	US liq gallons	US liq gallons	3.79	Liters
Weight					
Grams	0.0353	Ounces	Ounces	28.4	Grams
Kilograms	2.20	Pounds	Pounds	0.454	Kilograms
Metric tons	1.102	Short tons	Short tons	0.907	Metric tons
Metric tons	0.984	Long tons	Long tons	1.016	Metric tons

Appendix B

Doctrine and Training Literature

Users of this manual must understand the elements of doctrinal literature and their relationship to each other. The commonly used terms doctrine and tactics, techniques, and procedures (TTP) are interrelated and mutually supportive; yet, each term has its own usage, level of detail, and place in the hierarchy of doctrinal publications. Because there is a close interrelationship, publications overlap in what they provide to the user. Figure B-1 illustrates the relationship of these doctrinal terms and publications.

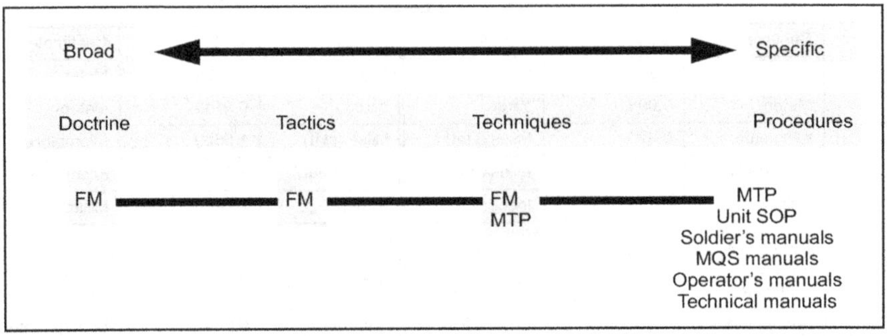

Figure B-1. Doctrinal and Training Literature

DOCTRINE

B-1. Doctrine is the body of fundamental principles by which the military guide their actions in support of national objectives. It is authoritative, but requires judgment in application. Doctrine provides the Army's fighting philosophy, establishes a common language, and structures the manner within which units plan and conduct combat operations.

B-2. Doctrine is not absolute; it is applied by the leader to meet the circumstances of the situation being faced. Doctrine provides for this flexibility with broad fundamentals and principles for conducting operations. Leaders well grounded in doctrine understand how to synchronize the elements of combat power and how to apply TTP using the assets available to accomplish the mission.

B-3. Doctrine is described in FMs. FM 3-0 defines the systems of war fighting practiced by the US Army. FMs 100-15 and 71-100 expand on this foundation. This manual describes the CS doctrine of the MP; it is the capstone manual for MP operations. It defines the role, operational requirements, functions, and C^2 relationships of MP units. This manual is not oriented on a specific organizational structure. Leaders at all levels apply this doctrine to their unique organizational and operational situations and use various tactics and techniques to accomplish the doctrine.

TACTICS

B-4. Tactics describe how the leader carries out doctrine (relating to the arrangement of forces for battle). Tactics include the—

- Employment of units in combat (to accomplish a mission assigned by the commander).
- Ordered arrangement and maneuver of units in relation to each other and to the enemy in order to use their full potential (a description of how the commander should arrange his forces and maneuver to accomplish a mission).

B-5. This is the part of doctrinal literature used for training in preparation for combat. Like doctrine, tactics presented in manuals are applied with judgment by the leader. The FMs prescribe "how to fight" and the mission training plans (MTPs) describe "what to train."

TECHNIQUES

B-6. Techniques are the methods of performing any act, especially the detailed methods used by troops or commanders in performing assigned tasks. Techniques describe the basic methods of using equipment and personnel. Techniques give details on how commanders actually carry out assignments. They improve the force's efficiency by ensuring the uniformity of action or by ensuring that the actions of various individuals and elements complement those of other individuals or elements.

B-7. The FMs and MTPs provide a description of techniques using the personnel and equipment available in actual organizations. These techniques show (at a more detailed level) how to use available assets to carry out tactics.

PROCEDURES

B-8. Procedures are the lowest level of detail. They address "how to" at the task level. Procedures are a particular course or mode of action that describe how to perform a certain task. Procedures include the—

- Standing methods used by units to accomplish tasks.
- Operating procedures for weapons and equipment.
- Crew drills.
- Staff actions and coordination requirements.
- Methods of target engagement by direct- and indirect-fire weapons.

B-9. Procedures are building blocks of individual- and collective-task accomplishment that serve as the foundation of tactics and techniques. Procedures are explained in unit standing operating procedures (SOPs), MTPs, soldier's manuals, operator's manuals, technical manuals (TMs), and similar publications.

Appendix C

Battlefield Workload Analysis

With minor adjustments in the capability block, the BWA can be used to determine MP requirements for the division. A BWA work sheet is shown in Figure C-1. A separate matrix is prepared for each EAC and corps MP brigade. Figure C-2, page C-3, is a sample of a BWA matrix for a 3-division corps MP CS company.

Mission/Function[1]	Capability[1]	Workload[2]	Rqmt	HN[18]	US Rqmt	Risk[19]	Mission Rqmt
Divisions supported[3]	1						
MMS	360 km MSR and ASR						
Refugees on control routes[4]	150,000 per day						
AS—corps rural terrain[5]	2,000 sq km						
AS—EAC rural terrain[5]	3,000 sq km						
AS—urban terrain[5]	800 sq km (urban)						
AS—USAF main opns base[5]	1,000 sq km/USAF MOB						
AS—convoy escort[6]	# tms per convoy	# convoys					
AS—ammo security[7]	1.3 modular ammo co						
	2 modular ammo co						
AS—pipeline security[8]	400 km						
AS—EAC pipeline co[8]	3						
AS—trains[9]	7						
AS—small critical sites[10]	12						
AS—major HQ security[11]	1						
AS—EAC port security[12]	0.66 of port						
AS—combat operations	When required, assets are diverted from other missions.						
L&O—incident response[13]	75,000 nondivision troops						
I/R—US prisoner detention[14]	700						
I/R—EPW escort/corps[15]	1,900 walking						
	2,500 by vehicle	Min: 1 plt					
	3,800 by train						
I/R—EPW holding/corps[16]	2,000	Min: 1 plt					
Other assigned missions[17]	Mission	Force size					
		Totals[20]					
Battalion HHD requirement = 0.199 x number of companies[21]							

Figure C-1. Sample BWA Work Sheet

NOTE: Workload ÷ capability = doctrinal requirement - HN offset = US requirement - acceptable risk = mission requirement

[1]Mission/capabilities are as stated in section I of TOEs 19477L000 and 19677L000. The unit capability for each mission is what the unit can accomplish when performing no other mission (for example, if it is only doing MSRs, the unit can cover 360 kilometers in a 24-hour period). Unless otherwise specified, the capabilities are the same at corps and EAC.

[2]Workload is the amount (MSR or area) or number of supported units/facilities.

[3]Corps support to each division is covered by allocating one MP company per supported division. Any additional support requirements should be based on a detailed division PM/MP mission analysis and included under other assigned missions.

[4]Unless specifically tasked, the Army normally relies on HN support to manage refugees. Redirecting refugees from MSRs and keeping them off of MSRs is part of (as is straggler control and TCP) the mission and MSR workload. When assigned the mission, the capability is for control of refugees on established refugee control routes that are in addition to the MSRs.

[5]Compute each square kilometer mission requirement separately. Do not compute the same terrain in more than one computation (for example, the area around the Air Force's main operating base (MOB) is not included with either rural or urban square kilometers).

[6]Convoy-escort capability is not stated in the TOE and is an AS function rather than an MMS function. Determine the average number of teams per mission and the average number of simultaneous missions. Multiply the number of teams required per mission by 0.028, then multiple the result by the number of simultaneous missions to determine the total requirement. State the number of teams per convoy and the number of convoys, then convert the force size to a 3-decimal-place company equivalent. Each additional team, squad, and platoon equivalent equals 0.028, 0.083, and 0.25 respectively.

[7]Security for modular ammo companies (ASPs) supporting divisions is three platoons and is normally conducted directly behind the division (corps forward). However, METT-TC may require the ammo units to move into the DRA. In this case, corps units should be attached to the division for the duration. Security for corps storage areas (CSAs) and theater storage areas (TSAs) is one MP company per two ammo companies. Security is provided on an area-support basis in the form of screening, with limited fixed posts.

[8]The EAC petroleum terminal/pipeline operating company manages the EAC pipeline that is normally laid adjacent to MSRs. Pipeline security is inherently provided by the MP operating on the MSRs. Unless the pipeline is not near the MSR, MP support is only allocated against the EAC petroleum terminal/pipeline operating company's bulk storage facilities. Security is provided on an area-support basis in the form of screening, with limited fixed posts.

[9]Train security is normally an EAC mission. This capability equates to simultaneously providing five MP teams to each of seven trains. Actual workloads and requirements are computed in the same manner as a convoy escort.

[10]Small critical facilities are normally defined as key MSR bridges and tunnels and communication nodes that may require additional security.

[11]The major headquarters are corps, COSCOM, TSC, unified command and, if directed, the JTF. The ASCC depends on the MP to provide security for two CPs. The ASCC requirement is an additional three platoons to support a second CP.

[12]The requirement for one seaport is 1.5 companies.

[13]Teams provide incident response, crime-scene security, and traffic control. They also assist L&O detachment teams and CID investigators.

[14]For deliberate planning in support of an OPLAN, personnel to be confined will normally be held by either the unit or the EAC I/R MP battalion performing the confinement mission. This does not preclude a requirement for the echelon PM to operate a field detention facility until prisoners are released back to their units or evacuated to the EAC confinement facility.

[15]The primary mode is vehicle (back-haul doctrine). Regardless of the mode, a minimum of one platoon (0.25 requirement) per supported division is allocated for EPW escort from the division central collection point (DCCP) to the corps holding area (CHA). Because the actual number of EPWs is unknown, coupled with no time/distance study for each supported division, one platoon per division provides a minimum capability for continuous evacuation within the stated doctrinal timeline (evacuate to the CHA within 12 hours of arrival at the DCCP).

[16]At a minimum, one platoon (0.25 requirement) is allocated to operate the CHA (provides the capability to hold 500 EPWs at any given time).

[17]Add any additional continuous missions assigned by the echelon commander not covered above. Examples include VIP escort/security for other than the CG and the deputy commanding general (DCG) of major headquarters, screening exposed corps flanks, and increasing teams per mission/function which decreases doctrinal capability (for example, permanent TCPs or checkpoints). Each additional team, squad, or platoon equivalent equals 0.028, 0.083, and 0.25 respectively.

Figure C-1. Sample BWA Work Sheet (continued)

[18]For each mission/function, any workload either fully or partially accomplished by the HN is subtracted from the doctrinal mission requirement. Enter the MP company equivalent offset and subtract from the doctrinal requirements to determine the US requirement.

[19]For each mission/function, if the operational/MP commander will accept mission risk, subtract the MP company equivalent to determine the mission requirement.

[20]Add columns to determine the total doctrinal company requirement, HN offsets, US requirement, acceptable risk, and mission requirement to support the operation/AO. Round the requirement to the nearest whole number (0.4 or fewer rounds down, 0.5 or greater round up).

[21]Determine battalion HHD requirements by multiplying the company requirements by 0.199 and rounding.

Figure C-1. Sample BWA Work Sheet (continued)

Mission/Function	Capability	Workload	Rqmt	HN	US Rqmt	Risk	Mission Rqmt
Divisions supported	1	3	3.000		3.000		3.000
MMS	360 km MSR and ASR	1,050	2.917		2.917		2.917
Refugees on control routes	150,000 per day	25,000	0.167	0.167	0.000		0.000
AS—corps rural terrain	2,000 sq km	6,500	3.250		3.250	0.500	2.750
AS—EAC rural terrain	3,000 sq km						
AS—urban terrain	800 sq km (urban)	1,500	1.875		1.875	0.500	1.625
AS—USAF main opns base	1,000 sq km/USAF MOB	1,000	1.000		1.000		1.000
AS—convoy escort	5 tms per convoy	4	0.560		0.560		0.560
AS—ammo security	1.3 modular ammo co	3	2.308		2.308		2.308
	2 modular ammo co	1	0.500		0.500		0.500
AS—pipeline security	400 km						
AS—EAC pipeline co	3						
AS—trains	7						
AS—small critical sites	12	9	0.750		0.750	0.250	0.500
AS—major HQ security	1	2	2.000		2.000		2.000
AS—EAC port security	0.66 of port						
AS—combat operations	When required, assets are diverted from other missions.						
L&O—incident response	75,000 nondivision troops	45,000	0.600		0.600	0.300	0.300
I/R—US prisoner detention	700						
I/R—EPW escort/corps	1,900 walking						
	2,500 by vehicle	Min: 1 plt/div	0.750		0.750		0.750
	3,800 by train						
I/R—EPW holding/corps	2,000	Min: 1 plt/	0.250		0.250		0.250
Other assigned missions	PSD for Corps CofS	1 sqd	0.083		0.083		0.083
	Flank screen	1 plt	0.250		0.250		0.250
		Totals	20	0.177	20	1.3	19
Battalion HHD requirement = 0.199 x number of companies			4		4		4

NOTE: Workload ÷ capability = doctrinal requirement - HN offset = US requirement - acceptable risk = mission requirement

Figure C-2. Sample BWA Matrix

Appendix D

Automated Information Systems in Support of Battle Command

As the MP Corps continues to advance into the twenty-first century, it must continue to pursue and stay in tune with the Army's initiative to acquire advanced technologies and operational concepts. This will give MP soldiers on the battlefield an information advantage that is critical for successful performance of their functions. The following information opens a window into the near future and explains some of the automation initiatives that will enhance the Army's and the MP Corps's soldiers.

ARMY INFORMATION SYSTEMS

D-1. The Global Command and Control System (GCCS) is the midterm solution and the bridge to the concepts outlined in the concept of command, control, communications, computers, and intelligence for the warrior (C4IFTW). The GCCS's core consists of the basic functions required by the war fighter to plan, execute, and manage military operations. The Army's selected C^2 system is the Army Battle Command System (ABCS).

D-2. The ABCS integrates the C^2 systems found at each echelon—from the ground-force component commander at the theater or JTF level to the individual soldier or the weapons platform. Whether ARFOR are deployed for land combat or are conducting peace operations, providing humanitarian assistance, or giving aid to civil authorities, the ABCS supports the mission by integrating the battle-space automation systems and communications that functionally link strategic and tactical headquarters. The ABCS is interoperable with joint and multinational C^2 systems at upper echelons across the full range of C^2 functionality, and it is vertically and horizontally integrated at the tactical and operational levels. The ABCS has three major components—

- The Global Command and Control System—Army (GCCS-A).
- The Army Tactical Command and Control System (ATCCS).
- The FBCB2.

GLOBAL COMMAND AND CONTROL SYSTEM—ARMY

D-3. The GCCS-A is the Army's component of the joint GCCS. It was built from application programs developed by the Army Worldwide Military Command and Control System [WMCCS] Information System (AWIS), namely the—

- Strategic Theater Command and Control System (STCCS).

- Theater Automated Command and Control Information Management System (TACCIMS).
- EAC portion of the Combat Service Support Control System (CSSCS) (CSSCS/EAC).

D-4. The GCCS-A is required in the MP LRP, S1, S3/NBC, and S4 sections within the MP brigade at EAC; in the Assistant Chief of Staff, G1 (Personnel) (G1), the G3, and the Assistant Chief of Staff, G4 (Logistics) (G4) sections within the MP command (I/R); and in the S1, S3, and S4 sections within the MP brigade (I/R).

ARMY TACTICAL COMMAND AND CONTROL SYSTEM

D-5. The ATCCS embodies the commander and his staff and the doctrine, procedures, and tools used to command and control forces on the tactical battlefield. The ATCCS's functional subsystems provide tools to support the C^2 process across the spectrum of low- to high-intensity conflicts worldwide; for MOOTW; and for combat, CS, and CSS functional areas at corps-and-below units. The ATCCS provides a common set of C^2 tools and functions for commanders that extends the ATCCS to the battalion level. The component's ATCCS's are the—

- Maneuver Control System (MCS).
- All-Source Analysis System (ASAS).
- Advanced Field-Artillery Tactical Data System (AFATDS).
- Air- and Missile-Defense Planning and Control System (AMDPCS).
- CSSCS.

D-6. The FBCB2 provides situational awareness and C^2 to the lowest tactical echelons. It facilitates a seamless flow of battle-command information and situation-awareness information across the battle space at brigade and below on Army platforms.

FORCE XXI BATTLE COMMAND—BRIGADE AND BELOW

D-7. The FBCB2 system, a battle-command information system, provides the technology to complete the ABCS information flow process from brigade to platform and across all platforms within the brigade TF. The FBCB2 enables commanders to remotely operate and maintain ABCS database connectivity (regardless of the command vehicle) and to digitally control and monitor their subordinate units' statuses and positions. Interoperable FBCB2 systems will enhance the battle commander's ability to have a detailed picture of the battlefield (situational awareness), to maximize force survivability, and to facilitate battlefield synchronization to reduce or destroy the enemy's capability to fight. The FBCB2 relies on integrating, processing, and moving information in a digital form. To support projections of future conflicts with an accompanying increase in the tempo of battle, the FBCB2 must be able to obtain information contained in databases from higher, lower, and adjacent elements without formally requesting that information through command channels. The war fighter requires the ability to pull data from all battlefield-system databases to keep current with the rapid changes caused by the dynamics of battle. The FBCB2 is required for divisional MP company commanders, PMs, platoon leaders, teams, and squads.

MANEUVER CONTROL SYSTEM

D-8. The MCS is the ABCS's horizontal and vertical integrator. The MCS will enable commanders and their staffs to collect, coordinate, and act swiftly on near-real-time battlefield information. It will quickly, accurately, and automatically move tactical information on the battlefield. This allows a commander to readily access information and display current situation reports (SITREPs) that assess enemy strength and movement as well as the status of friendly forces. The MCS will then aid the battle staff in rapidly disseminating the commander's orders. It is intended to provide two major functional capabilities—maneuver functional-area control and force-level information. The MCS is the ATCCS's functional subsystem that—

- Supports the combined-arms force commander directly by providing automated support for planning, coordinating, controlling, and using maneuver functional-area assets and tasks.
- Coordinates and synchronizes supporting arms in conducting operational planning, field operations, and training.

D-9. The MCS is required for division PM operations sections, company CPs at echelons corps and below (ECB), MP battalion commanders, S3 sections at corps, MP brigade commanders, and corps MP LRP and S3 sections.

ALL-SOURCE ANALYSIS SYSTEM—REMOTE WORK STATION

D-10. The ASAS—remote workstation (RWS) subsystem will operate within the existing intelligence and electronic warfare (IEW) battlefield operating system (BOS) and ABCS architectures to provide automation support to G2/S2 intelligence staff elements and other designated intelligence organizations from the maneuver battalion through EAC. It will operate at the security level appropriate to its echelon, mission, and functions. It assists the G2/S2 in providing IEW support to the commander and other staff elements in planning, coordinating, and conducting current and future tactical operations. It interfaces with the sensitive compartmented information (SCI) analysis and control element (ACE) subsystem to provide operational guidance and to exchange IEW information and products. It also operates as the technical control portion of the IEW C^2 node of the ABCS to provide current IEW and enemy-situation information to the force-level information database for access and use by ABCS users. Two RWSs are also provided to the brigade analysis control team (ACT) of the DS MI company to process data input from DS collection systems, to filter intelligence data from broadcast systems, and to pass tactically relevant enemy-situation information or situation-development data to the brigade S2's RWS. The ASAS-RWS is required in the S2 section of the CID group, the S2/S3 section of the CID battalion, the G2 section of the MP command (I/R), the S2 section of the MP battalion (I/R), and the S2 section of the MP battalion and brigade at all echelons.

COMBAT SERVICE SUPPORT CONTROL SYSTEM

D-11. The CSSCS will function as an integral part of the ATTCS. It is designated as an information system with the capability to rapidly collect, analyze, and disseminate logistical, medical, and personnel information to accelerate and improve decision making by CSS and force commanders. It also enables the rapid exchange of C^2 information with other battlefield functional areas (BFAs) through automated interfaces. The system will process unclassified through secret information. The CSSCS will be employed in all divisions, corps, EAC, separate brigades, and armored cavalry regiments. The MP require CSSCS functionality in all S1 and S4 sections in battalion or higher organizations except where the GCCS-A is required as stated above.

TRANSPORTATION COORDINATOR AUTOMATED MOVEMENT SERVER

D-12. The Transportation Coordinator Automated Movement Server (TC-AIMS II) is used by transportation agents and deploying units of each service to automate the processes of planning, organizing, coordinating, and controlling deployment and redeployment operations worldwide during peace as well as contingencies. The TC AIMS II will link all DOD-component unit movements with the installation transportation officer/traffic-management officer (ITO/TMO). This will provide the capability to plan and execute deployments rapidly to meet CINC-required arrival and departure dates. The TC AIMS II will also provide the ITO's/TMO's day-to-day traffic management functions and the theater transportation managers with the capabilities to use theater common lift assets effectively. The TC AIMS II will also allow CSS organizations and supporting commands to sustain the force and perform effective movement control, coordination, distribution, and sustainment once in theater. This is also defined as transportation movement and support of DOD personnel and cargo during all phases of military operations in all environments (including RSOI and battlefield operations). The TC AIMS II's functionality is required at all division PM operations and S3 sections within MP battalions and brigades to facilitate MSR regulation enforcement. Additionally, MP companies, battalions, brigades, and commands require the TC AIMS II to plan, organize, coordinate, and control deployments and redeployments.

GLOBAL COMBAT SUPPORT SYSTEM—ARMY

D-13. The GCSS-A is an integrated logistical support system with accompanying hardware that will connect the service's supply, finance, transportation, maintenance, and personnel databases. Under this system, the Army will integrate 13 Legacy tactical logistics systems and will eventually interface with the service's wholesale logistics systems as well as the consolidated logistics systems of the other services and the DOD. The system is critical for modernizing automation, for reducing logistical footprints in theaters, and for expanding war fighters' capabilities. The system will also replace the Standard Army Management Information System (STAMIS). The GCSS-A is required at all MP division company CPs and MP battalion and brigade S4 sections. Additionally, it is required at all supply functions within MP detachments that support I/R missions.

MILITARY POLICE AUTOMATED SYSTEMS

D-14. In an effort to maintain an information advantage and to acquire advance technologies and operational concepts, the MP Corps has acquired the following systems:

- The Military Police Management Information System (MPMIS) and its subsystems
- The CID's automated systems and their subsystems.

MILITARY POLICE MANAGEMENT INFORMATION SYSTEM

D-15. The MPMIS is a DA-mandated reporting system that does not interface with the ABCS. The MPMIS provides automated tools for MP records management and reporting. It reduces the administrative burden on the PM's staff. It increases the awareness of problem areas by managing crime information. The MPMIS consists of the following elements:

- **Offense Reporting Systems Windows (ORS-2W).** The ORS-2W provides automation support to track and report complaints, criminal incidents, and traffic incidents. It incorporates the CRC, the federally mandated National Incident-Based Reporting System (NIBRS), the DOD-mandated Defense Incident-Based Reporting System (DIBRS), the military community (MILCOM), and the installation's data requirements. The ORS-2W's functionality is required at PM activities at all echelons of command within the Army.
- **Correctional Reporting System (CRS-3).** The CRS-3 provides automation support to track and report inmates and corrections staff at all Army-controlled correctional facilities.
- **Security Management System (SMS).** The SMS provides automated support to assist physical-security inspectors and force-protection officers in reporting inspections and surveys. It tracks intrusion-detection devices and MWD training and health requirements.
- **Vehicle Registration System (VRS).** The VRS provides installations with an automated means to track vehicle, weapon, and bicycle registrations.
- **Prisoner-of-War Information System (PWIS-3).** The PWIS-3 provides the automated means of tracking an EPW's status from capture through release. This system is required at all EPW management activities.

CID AUTOMATED SYSTEMS

D-16. The Headquarters (HQ) USACIDC mandates CID automated systems. Similar to the MPMIS, it does not interface with the ABCS. The system facilitates data-collection efforts within the command. The CID's automated systems consist of the—

- **Automated Criminal Investigative Reporting System (ACIRS).** The ACIRS supports management and data-collection functions at the CRC and HQ USACIDC. It supports the following:
  - Maintaining investigative data.
  - Conducting agent investigative activities.
  - Reporting investigative data.
  - Managing the investigative processes.
  - Conducting criminal-activity analyses.

- **CRIMINTEL.** Data sources for CRIMINTEL include MP reports, CRIMINTEL reports and bulletins, reports of investigation, crime-prevention surveys, case files, collection folders, raw-data folders, significant data files, and target-analysis files. The CRIMINTEL interfaces with the ACIRS, but it does not share CRIMINTEL products.

Appendix E

Police Information Assessment Process

The PIAP differs somewhat from the tactical IPB. While the IPB allows the maneuver commander to see the effects that a variety of factors may have on his forces, the PIAP looks beyond this approach to determine how other relevant information may impact his forces. This information may be obtained from police or criminal actions or incidents encountered during the performance of MP functions.

IMPLEMENTATION

E-1. The PIAP is not a substitute for the IPB. MP leaders must continue to rely on the IPB and to use its estimates as a starting point for the PIAP. However, the PIAP is a dynamic and continuous cycle that complements the IPB (Figure E-1). The following paragraphs represent the doctrinal approach to implementing the PIAP. These steps provide the basic foundation to the process. They are not exclusive; the successful development of the PIAP will depend on the MP leader's ability to apply the process to his specific environment, METT-TC, and the commander's priorities.

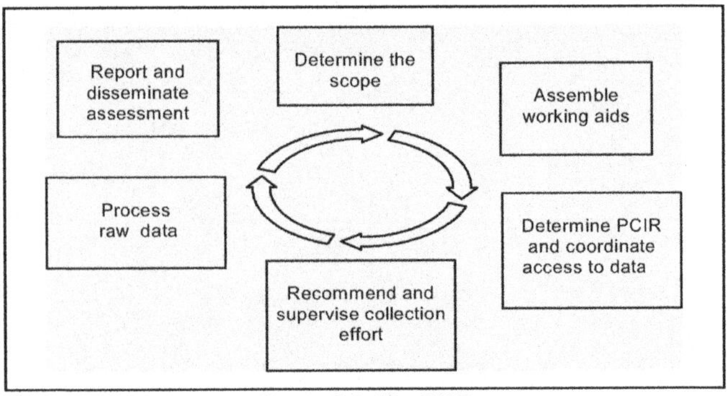

Figure E-1. The PIAP

DETERMINE THE SCOPE OF THE PIAP

E-2. While determining the scope of the PIAP might sound simple, it is one of the most crucial elements of the entire process. It entails identifying what you want to be the end result of your PIAP. To help determine the scope of the PIAP, MP leaders must perform the following subtasks:

- **Conduct a mission analysis.** A detailed mission analysis will ensure that all available resources are focused toward one goal. The more specific the mission is, the more focused and productive the effort. This involves more than just looking at the mission statement, the commander's intent, and the concept of operations. It involves translating the mission and situation into easily understood, manageable goals. The following are examples of these missions:
  - Identifying criminal threats from local national to US Army logistics operations within the port of Mogadishu, Somalia.
  - Determining the impact of gang violence on US forces operating in the vicinity of Camp Bondsteel, Kosovo.
  - Determining how rioters' violence will affect force protection before deploying forces to Los Angeles, California.
  - Determining the status of public service agencies (such as the police, the fire department, and the emergency medical services [EMS]) in Homestead, Florida, after the city was struck by Hurricane Andrew.

NOTE: As you can see from the examples, these missions are quite specific. In all likelihood, it would be necessary to establish several different PIAP missions to adequately protect Army elements operating in the MP's AOR.

- **Prioritize missions.** MP leaders must prioritize each PIAP mission since MP resources will probably be limited. The commander's intent is the key factor for determining priorities. When the commander's most important mission has been determined, MP leaders must decide how they can contribute to the overall success of that mission.
- **Determine the AO.** Generally, the AO is a geographical area (including the airspace above) usually defined by lateral, forward, and rear boundaries assigned to a commander. MP leaders must know where major units are located and how the characteristics of the location may impact ongoing operations. For example, units in direct contact with the enemy or isolated from the local population are less likely to be the target of a PIAP mission than those units adjacent to great population centers or in rest and relaxation centers.
- **Determine key terrain.** Key terrain is any locality or area that affords a marked advantage to the retaining side.
- **Determine the AOI.** An AOI is the area of concern to the commander. This includes the area of influence and the areas adjacent thereto and extending into enemy territory and to the objectives of current or planned operations. A thorough understanding of the AOI may help predict potential threat actions. For example, how far are local agitators and ringleaders willing to travel to interfere with US operations in Pristina, Kosovo? Would they go to other US compounds?
- **Determine required information products/sources.** The desired end-state product must be decided. Should it be a report? An updated SITREP? An upgrade in force-protection measures? Or a combination of products? It should also be determined who can provide the best

information (the CID; MI; the local police chief; or joint, interagency, or multinational agencies).

ASSEMBLE WORKING AIDS

E-3. During this step, MP leaders assemble the necessary graphic and nongraphic aids and information that will help them accomplish their mission. The CID and the MP may form an analysis team to track and analyze police information that might assist tactical operations or population-control operations. Working aids include, but are not limited to—

- **Maps.** Besides the standard military maps that depict the military AOs, the MP must attempt to procure indigenous maps of AOs. Detailed city and street maps will prove to be invaluable even if they are in a foreign language.
- **Language aids.** Phrase books and translation dictionaries cannot take the place of intensive language training, but they are better than nothing. These are essential even if the MP have assigned interpreters.
- **Open-source information.** This can cover an extensive range of invaluable information, but it should be available from a variety of sources. Examples include—
  - Demographic information on the HN population.
  - Societal information on the HN.
  - The locale and disposition of HN police forces.
  - Historical crime data of the area.
  - Environmental information (terrain and weather factors).
  - Local newspaper articles.
  - Internet sources on all of the above.

NOTE: Sources for this type of information can come from intelligence estimates provided by the higher headquarters' S2/G2, other military units, or the HN.

- **OPSEC information.** This information includes the disposition of US forces (to include MP, MI, CID, and other services' investigating units) deploying or being used within the AO.
- **Other working aids.** These may include items such as computers, printers, boards, and acetate that will be necessary depending on the missions the MP are undertaking.

DETERMINE INFORMATION REQUIREMENTS AND COORDINATE ACCESS TO DATA

E-4. Once the mission analysis is completed, the next step is to determine if there are information gaps. An information gap is a missing piece of information that is critical to the analysis. MP leaders must identify and prioritize the gaps using the time, the available resources, and the commander's intent. Once an information gap is identified, it becomes PCIR. Examples of PCIR are as follows:

- Where is the nearest police station, and how did they handle a certain problem?

- What is the composition of the two gangs operating in our AO, and how can they influence US operations?
- Why is river-crossing point Bravo a danger to US crossing forces? Why are local rioters disrupting only US operations on the east bank?

E-5. The next step is to determine if PCIR can be answered with data already available. To do this, the available military and nonmilitary sources must be identified. Some sources may include—

- **USACIDC.** The USACIDC is responsible for programs such as combating terrorism, CRIMINTEL, personal-security assessment, crime analysis, and LOGSEC threat assessment. Some of the information gathered by the USACIDC (although releasable under existing controls and restraints) may help with the PIAP mission. In fact, USACIDC special agents may already be working in the AO and may be available to help the MP with the collection effort.
- **MI element.** Like the USACIDC, the MI element operating in the AO may have come across a piece of information (police or criminal) that may help accomplish the PIAP mission. The MP must make daily contact with the S2/G2 and coordinate access to invaluable information. As with the USACIDC, available intelligence information may only be releasable under strict controls and restrictions imposed by the commander.
- **Other MP units.** Coordination with an adjacent MP unit may prove to be beneficial, especially if the other MP unit has conducted a similar PIAP mission or has conducted operations in the AOI.
- **HN law-enforcement agencies.** Local law-enforcement agencies will provide the essential populace knowledge otherwise not available to US forces. Chiefs of police, sheriffs, and other key community leaders may help MP leaders fill the information need of the PIAP mission.
- **Joint, interagency, and multinational forces.** Close and effective liaison with these elements will result in quick access to information that can produce the same benefits as the agencies listed above.

NOTE: If, after consulting with all available sources, the PCIR is not resolved, the MP leader must decide the best collection effort to solve the PCIR.

RECOMMEND AND SUPERVISE THE POLICE/CRIMINAL COLLECTION EFFORT

E-6. The collection effort is the means by which specific PCIR will be met. This implies developing a collection strategy, tasking specific collectors, and supervising the collection effort.

- **Develop a collection strategy.** After a through study of the availability, capability, and disposition of the potential collecting resources, MP leaders select which asset is better suitable to perform the mission. Are organic MP the best collectors for this mission? Or do we need to request support from USACIDC units? What unique or organic capabilities do each bring to the effort? Which MP function is the best? Do we conduct MMS, AS, or L&O operations? Part of the collection strategy includes coordinating with the S2/G2, the SJA, the

CID, and other agencies before launching the collection effort. This coordination will eliminate duplication of effort, interference with an ongoing effort, or stepping out of legal limits.

- **Task or request specific collectors.** The missions that will be tasked to respective collectors must be determined. Information collectors can be tasked with more than one mission at a time. However, it is imperative that their tasks be prioritized based on mission requirements and time available. The appropriate tasking or request chain must be used to request an MP team, a platoon, or USACIDC special agents.
- **Supervise the collection effort.** The collectors must be provided with reporting guidelines. How often should they report? Should they report "no information?" Specific reporting instructions should be provided, including how, when, and where they report. A collection tasking chart should also be used (Figure E-2). As more and more collectors are being used, it becomes important to track their missions, capabilities, and success. A brief note on the reporting instructions of each collector should be included.

PIAP Missions	Unit Tasked	Echelon Employed	POC	Collection Strategy Employed	Date Collection Started	Location	Date Collection Ended
PCIR 1							
PCIR 2							
PCIR 3							
PCIR 4							

Figure E-2. Collection Tasking Chart

PROCESS POLICE/CRIMINAL RAW DATA

E-7. This processing step converts raw data into police intelligence through analysis. The key to processing the data is to understand it. For example, as collectors report raw data it may often appear to be meaningless information. Rarely will one collector receive all of the necessary information to answer the PCIR at one time. Instead, a variety of information must be fitted together to form a coherent, even if incomplete, picture.

E-8. Every piece of data is important. This is the cardinal rule of the entire process. Until the data is analyzed for reliability and compared with other data, all reports must be treated equally. This is important to remember, for the trend will be to use data that either fits your expectations or gives the most complete picture.

E-9. A report of "no activity" can be as important as reports with information. If the collector is capable of collecting the information and is actively trying to collect, reports of "no activity" can be critical. First, if there is no indication of a change, they can be used to reprioritize PIAP missions. Second, they can be used to retask the collectors to other areas. It is important to check the mission's original intent periodically to ensure that it is still valid and merits further efforts.

E-10. To determine the reliability of the data, the MP leader must evaluate the source. Does the information come from the police chief, local youth, or other military forces? One of the best ways to determine the reliability of the information is to have more than one collector reporting on the same PCIR. If more than one collector reports the same information, it can normally be presumed to be accurate.

E-11. Information can be analyzed in a variety of ways—chronologically, geographically, by the impact on the force, by the type of activity or the modus operandi, or by the association of participants. The MP leader must evaluate the information after it is analyzed to determine whether the information collected answers the PCIR or whether it helps accomplish the PIAP mission. If the information does not answer these questions, he must decide whether it is necessary to task other collectors, retask the same collectors, or wait until the situation develops.

REPORT AND/OR DISSEMINATE POLICE AND CRIMINAL ASSESSMENT

E-12. Reporting and disseminating is the most important aspect of the PIAP. Collection and analysis is valueless unless the data is communicated to commanders who can best use the information.

E-13. Reporting requirements must be established based on the unit's SOP and other established processes. MP leaders must decide if the information collected must be sent using an established priority format or if it can wait to be sent using SITREPs. Does the information collected affect the force-protection measures or does it immediately impact the tactical scenario? Do we need to alert units operating in the vicinity of AA Fox to avoid sector 2? Additionally, MP leaders must ensure that any information released follows existing constraints and regulations.

E-14. It is also advantageous to provide feedback to the collectors. Collectors often receive no feedback on their performance. Informing them as to which part of their information was vital and beneficial can often increase their productivity in future operations.

SAMPLE SCENARIO

E-15. The following example helps illustrate the implementation of all six steps of the PIAP:

E-16. While deployed in support of Operation Control Chaos, US forces assigned to TF Blue Thunder are tasked with stability and support operations in the city of Corbina. This city, although currently implementing a peace treaty, was the site of extreme civil unrest. Even though the overall situation is not as chaotic as it was before US intervention, there are some sectors that are still in turmoil. Common crimes (such as robberies, assaults, looting, and vandalism) still occur daily and there is fear that this condition may spread out of control throughout the city. The TF commander knows that in order to bring peace to the entire city, he must attempt to stabilize the situation in some of these bad sectors. He decides that the best way to deal with this situation is to assign a sector (sector 51) to an MP battalion in an attempt to control the situation. If the MP battalion succeeds, he may use the same concept throughout the city. He is aware of the five MP functions and how they can help bring this sector under some control. Under this framework, the TF commander issues the following mission: *On order, the MP battalion will conduct stability and support operations in the vicinity of sector 51, the city of Corbina, in support of TF Blue Thunder.* Upon receipt of this mission, the battalion commander gathered his staff and initiated the planning process. Before the MP staff could initiate the PIAP, they made contact with the S2/G2 and received an updated IPB estimate. Now that the IPB estimates are on-hand, the staff initiates the PIAP.

DETERMINE THE SCOPE OF THE PIAP

E-17. The staff performs the following steps to determine the scope of the PIAP:

- **Conduct a mission analysis.** The staff looks at the given mission, the higher echelon commander's intent, and the concept of operations. After considering all of these factors, they produce a restated mission—*On order, the battalion will conduct MP operations in sector 51 to determine the type of threat and to bring this sector under control in support of TF Blue Thunder.*
- **Prioritize PIAP missions.** Since there is only one mission, this subtask is not applicable.
- **Determine the AO and the AOI.** The staff determines that the AO is the entire sector 51 and that the AOIs are the sectors adjacent to 51.
- **Determine key terrain.** The staff determines that the key terrains are the market place, the plaza, and the city park. They select these areas because most of the incidents occurred in these populated areas. Stabilization in these key areas may be advantageous to mission success.
- **Determine required information products/sources.** The end result is the stabilization of sector 51. Minimize the crime incidents and restore L&O. Information should come from the local police chief, local community leaders, civil affairs, PSYOP, and the CID.

ASSEMBLE WORKING AIDS

E-18. The staff assembles the following working aids in support of the PIAP:

- **Maps.** The staff acquires local maps that detail the market place, the plaza, and the city park. Although the maps are in a foreign language, the staff is able to identify these three prominent features and surrounding areas.
- **Language aids.** In addition to dictionaries, the staff is able to get two translators from civil affairs.

- **Open-source information.** Open-source information includes a variety of encyclopedias and country books. Newspaper reports and historical crime data is available via the internet or through MI sources. The emphasis is in obtaining information on the history and composition of the local police force.
- **Classified intelligence.** The staff receives a briefing from the S2/G2 describing the operations of SOF units in the area.
- **Other working aids.** Computers, charts, templates, and acetate are available.

DETERMINE INFORMATION REQUIREMENTS AND COORDINATE ACCESS TO DATA

E-19. After a careful analysis of all available information, the following information gaps were turned into PCIR:

- Why is the police force not actively engaged in L&O?
- What is the major threat to civilians in each area?
- Are local gangs involved?
- Is their local judicial system in place?
- Where are prisoners housed?

E-20. The staff consults with the USACIDC, MI, other MP units, and other sources but determines that the information provided is not clear or sufficient. Since the PCIR are not resolved, the staff decides to recommend that the commander initiate a collection effort.

RECOMMEND AND SUPERVISE THE POLICE/CRIMINAL COLLECTION EFFORT

E-21. The staff performs the following steps to recommend and supervise the police/criminal collection effort:

- **Develop a collection strategy.** Upon coordinating with the S2/G2, the SJA, and the CID, the MP commander decides that the best collectors will come from within organic assets. He decides that he will employ the MP to go out and seek the needed information. The battalion commander will employ a company commander to interview the local police chief and an MP/interpreter will patrol around the designated AOI. The MP commander will need additional interpreters to help him with his collection effort.
- **Task or request specific collectors.** The MP staff requests an interpreter through the normal chain of command. The TF G3 assigns six interpreters for a period of 90 days.
- **Supervise the collection effort.** All collectors are instructed to report not later than (NLT) 1800 every day. The report will be in the form of an "end of day" outbrief to the S2 and the battalion commander. A collection tasking chart (Figure E-3) will also be completed.

PROCESS POLICE/CRIMINAL RAW DATA

E-22. After receiving all of the above information, the S2 concludes that the inability of the police force to do their job has had a domino effect on all areas of peace and order. This information is confirmed through other reports from information collected by the CID (through their own efforts) and the CA. The information also reveals that the gangs have intentions of attacking US forces if they interfere with their "business." The S2 determines that although the situation may seem to be chaotic, once the locals are able to trust the police forces, everything should return to normal.

REPORT AND/OR DISSEMINATE POLICE AND CRIMINAL ASSESSMENT

E-23. Since the collection effort identifies a potential threat to US forces, the commander decides to notify higher headquarters and increase his own force-protection measures. He requires MP elements patrolling in designated areas to be no less than a squad. If patrolling dismounted, a HMMWV or an ASV will trail the squad. The MP commander uses the information provided during the PIAP to develop his implementation plan. This plan includes joint patrols, training, and construction of a joint MP/local national police station.

	PIAP Mission: Bring Sector 51 Under Control							
PCIR	**Unit Tasked**	**Echelon Employ-ed**	**POC**	**Collection Strategy Employed**	**Date Collection Started**	**Location**	**Date Collection Ended**	**PCIR Result**
Why is the police force not actively engaged in L&O operations?	A Co	2 teams	CPT Smith	Direct interview	31 Jan 00	Sector 51, Police HQ	2 Feb 00	The police force does not know how to interact with the population.
What is the major threat to civilians in each area?	A Co	6 teams	1LT Thomas	Patrol and interview	31 Jan 00	Market place, plaza, and park	7 Feb 00	Muggings, robberies, and stabbings are the major threats to civilians.
Are local gangs involved?	A Co	2 teams	CPT Smith	Direct interview	31 Jan 00	Sector 51, Police HQ	2 Feb 00	Two rival gangs exist.
Is a local judicial system in place?	A Co	2 teams	1LT Jones	Direct interview	31 Jan 00	Courthouse	3 Feb 00	A judicial system exists, but the judge is afraid to sentence anyone.
Where are the prisoners housed?	A Co	4 teams	1LT Lucas	AS and MMS	31 Jan 00	Local prison	2 Feb 00	The local prison is abandoned.

SAMPLE

Figure E-3. Sample Collection Tasking Chart

Glossary

1LT	first lieutenant
A/L	administrative and logistics
AA	assembly area
AATF	air-assault task force
ABCS	Army Battle Command Systems
ACE	analysis and control element
ACIRS	Automated Criminal Investigative Reporting System
ACT	analysis control team
ADA	air-defense artillery
ADC	area damage control
AFATDS	Advanced Field-Artillery Tactical Data System
ALO	authorized level of organization
AMDPCS	Air- and Missile-Defense Planning and Control System
AO	area of operation
AOE	Army of Excellence
AOI	area of interest
AOR	area of responsibility
APOD	aerial port of debarkation
APOE	aerial port of embarkation

AR	armored
AR	Army regulation
ARFOR	Army forces
AS	area security
ASAS	All-Source Analysis System
ASCC	Army service component command
ASG	area support group
ASP	ammunition supply point
ASR	alternate supply route
ASV	armored security vehicle
ATCCS	Army Tactical Command and Control System
attn	attention
AWIS	Army WMCCS Information System
BCT	brigade combat team
bde	brigade
BFA	battlefield functional area
BLD	brigade liaison detachment
bn	battalion
BOS	battlefield operating system
BRT	brigade reconnaissance troop
BWA	battlefield workload analysis

C^2 command and control

C^3 command, control, and communications

C4I command, control, communications, computers, and intelligence

C4IFTW command, control, communications, computers, and intelligence for the warrior

CAS close air support

CASE corps-area support element

CCIU computer-crime investigative unit

cdr commander

CG commanding general

CHA corps holding area

CI civilian internee

CID Criminal Investigation Division

CIDR Criminal Investigation Division Regulation

CINC commander in chief

CIP Criminal Intelligence Program

cmd command

CMOC civil-military operations center

co company

COA course of action

CofS Chief of Staff

commo communications

COMMZ	communications zone
conf	confinement
CONUS	continental US
COSCOM	corps support command
CP	command post
CPT	captain
CRC	Crime Records Center
CRIMINTEL	criminal intelligence
CRS-3	Correctional Reporting System
CS	combat support
CSA	corps storage area
CSE	corps support element
CSG	corps support group
CSM	command sergeant major
CSS	combat service support
CSSCS	Combat Service Support Control System
cust	customs
CZ	combat zone
DA	Department of the Army
DC	District of Columbia
DCCP	division central collection point

DCG	deputy commanding general
DIBRS	Defense Incident-Based Reporting System
DISCOM	division support command
div	division
DLEA	domestic law-enforcement agency
DMMC	division materiel management center
DNA	deoxyribonucleic acid
DOD	Department of Defense
DOJ	Department of Justice
DOS	Department of State
DPM	deputy provost marshal
DRA	division rear area
DS	direct support
DSA	division support area
DSB	division support battalion
DSE	division support element
DSS	decisive, shaping, and sustaining
DZ	drop zone
EAC	echelons above corps
EAD	echelons above division
ECB	echelons corps and below

EG	escort guard
EMS	emergency medical services
EPW	enemy prisoner of war
ERP	engineer regulation point
ES	escort
FA	field artillery
FBCB2	Force XXI Battle Command—Brigade and Below
FBI	Federal Bureau of Investigation
Feb	February
FEBA	forward edge of the battle area
FIU	field investigative unit
FM	field manual
FOB	forward operating base
FOIA	Freedom of Information Act
FORSCOM	US Army Forces Command
FSB	forward-support battalion
FSC	forward-support company
G1	Assistant Chief of Staff, G1 (Personnel)
G2	Assistant Chief of Staff, G2 (Intelligence)
G3	Assistant Chief of Staff, G3 (Operations and Plans)
G4	Assistant Chief of Staff, G4 (Logistics)

G5	Assistant Chief of Staff, G5 (Civil Affairs)
G6	Assistant Chief of Staff, G6 (Communications)
GAC	ground-assault convoy
GCCS	Global Command and Control System
GCCS-A	Global Command and Control System—Army
GD	guard
GO	government organization
GS	general support
HHB	headquarters and headquarters battery
HHC	headquarters and headquarters company
HHD	headquarters and headquarters detachment
HMMWV	high-mobility, multipurpose, wheeled vehicle
HN	host nation
HQ	headquarters
HRP	high-risk personnel
HVA	high-value asset
I/R	internment and resettlement
IBCT	initial/interim brigade combat team
IC	information center
IEW	intelligence and electronic warfare
IG	inspector general

IHO	international humanitarian organization
IMETP	International Military Education and Training Program
IO	international organization
IPB	intelligence preparation of the battlefield
IRIC	internment/resettlement information center
ISB	intermediate staging base
ITO	installation transportation officer
J3	Operations Directorate
Jan	January
JFC	joint-force command
JOA	joint-operations area
JP	joint publication
JRA	joint rear area
JSOA	joint special operations area
JTF	joint task force
km	kilometer(s)
L&O	law and order
LID	light infantry division
liq	liquid
LOC	lines of communication
LOGSEC	logistics security

LRP long-range plan

LSTA logistics-security threat assessment

LVOSS Light-Vehicle Obscuration Smoke System

LZ landing zone

M/O marksman/observer

M/S mobility/survivability

MACOM major Army command

MAJ major

MCRP Marine Corps reference publication

MCS Maneuver Control System

MDMP military decision-making process

mech mechanized

MEDEVAC medical evacuation

METL mission-essential task list

METT-TC mission, enemy, terrain, troops, time available, and civilian considerations

MEVA mission-essential or vulnerable area

MFPU major fraud procurement unit

MI military intelligence

MILCOM military community

min minimum

MKT mobile kitchen trailer

MLRS Multiple-Launch Rocket System

MMS maneuver and mobility support

MOB main operating base

MOOTW military operations other than war

MP military police

MPBTF military police brigade task force

MPMIS Military Police Management Information System

MQS military qualification standards

MSR main supply route

MTOE modified table of organization and equipment

MTP mission training plan

MTT mobile training team

MWD military working dog

NATO North Atlantic Treaty Organization

naut nautical

NBC nuclear, biological, and chemical

NCO noncommissioned officer

NEO noncombatant evacuation operation

NG national guard

NGO nongovernmental organization

NIBRS National Incident-Based Reporting System

NLT not later than

NLT nonlethal weapon

NWP Naval warfare publication

OCONUS outside the continental US

OPCON operational control

OPLAN operation plan

opns operations

OPSEC operations security

ORS-2W Offense Reporting Systems Windows

PASE port-area support element

PCIR police/criminal information requirements

PDF Panamanian Defense Force

PEO peace-enforcement operation

PI police intelligence

PIAP police information assessment process

PIO police intelligence operations

PIR priority intelligence requirements

PKO peacekeeping operation

plt platoon

PM provost marshal

POC point of contact

POD port of debarkation

POE port of embarkation

POL petroleum, oil, and lubricants

PRC populace and resource control

PSD protective security detail

PSU protective-service unit

PSVA personal-security vulnerability assessment

PSYOP psychological operations

pub publication

PWIS-3 Prisoner-of-War Information System

PZ pickup zone

R&S reconnaissance and surveillance

RAOC rear-area operations center

RC reserve component

ROC rear operations center

ROE rules of engagement

rqmt requirement

RSOI reception, staging, onward movement, and integration

RSTA reconnaissance, surveillance, and target acquisition

RWS remote workstation

S1 Adjutant (US Army)

S2 Intelligence Officer (US Army)

S3 Operations and Training Officer (US Army)

S4 Supply Officer (US Army)

S5 Civil-Affairs Officer (US Army)

S6 Communications Officer (US Army)

SALUTE size, activity, location, unit, time, and equipment

SAW squad automatic weapon

SCI sensitive compartmented information

sec section

SEP separate

SFC sergeant first class

SITREP situation report

SJA staff judge advocate

SMS Security Management System

SOC security-operations cell

SOF special-operations forces

SOFA status of forces agreement

SOI signal operating instructions

SOP standing operating procedures

SPOD seaport of debarkation

SPOE seaport of embarkation

SPOTREP	spot report
spt	support
sq	square
sqd	squad
SRT	special-reaction team
STAMIS	Standard Army Management Information System
stat	statute
STCCS	Strategic Theater Command and Control System
TA	theater Army
TAA	tactical assembly area
TAACOM	Theater Army Area Command
TAC	tactical
TACCIMS	Theater Automated Command and Control Information Management System
TAI	traffic accident investigation
TASE	TSC-area support element
TC-AIMS	Transportation Coordinator Automated Movement Server
TCF	tactical combat force
TCP	traffic-control post
TF	task force
THREATCON	threat condition
TM	technical manual

TMO	traffic-management officer
tms	teams
TO	theater of operations
TOE	tables of organization and equipment
TRADOC	US Army Training and Doctrine Command
TRP	traffic-regulation plan
TSA	theater storage area
TSC	theater support command
TSE	tactical support element
TTP	tactics, techniques, and procedures
UMT	unit ministry team
UN	United Nations
US	United States
USACIDC	US Army Criminal Investigation Command
USACIL	US Army Criminal Investigations Laboratory
USAF	US Air Force
USAMPS	US Army Military Police School
USC	US Code
USCG	United States Coast Guard
USCS	US Customs Service
USDA	US Department of Agriculture

VIP very important person

VRS Vehicle Registration System

WMCCS Worldwide Military Command and Control System

Bibliography

AR 15-6. *Procedures for Investigating Officers and Boards of Officers.* 11 May 1988.

AR 190-12. *Military Police Working Dogs.* 30 September 1993.

AR 190-45. *Law Enforcement Reporting.* 30 September 1988.

AR 190-47. *The Army Corrections System.* 15 August 1996.

AR 195-2. *Criminal Investigation Activities.* 30 October 1985.

AR 525-13. *Antiterrorism / Force Protection (AT / FP): Security of Personnel, Information, and Critical Resources.* 10 September 1998.

CIDR 195-1. *Special Agent Operations Manual.* October 1994.

DA Form 2028. *Recommended Changes to Publications and Blank Forms.* 1 February 1974.

FM 3-0. *Operations.* To be published within the next 6 months.

FM 19-1. *Military Police Support for the AirLand Battle.* 23 May 1988.

FM 19-4. *Military Police Battlefield Circulation Control, Area Security, and Enemy Prisoner of War Operations.* 7 May 1993.

FM 19-10. *The Military Police Law and Order Operations.* 30 September 1987.

FM 19-40. *Enemy Prisoners of War, Civilian Internees and Detained Persons.* 27 February 1976.

FM 34-2. *Collection Management and Synchronization Planning.* 8 March 1994.

FM 34-130. *Intelligence Preparation of the Battlefield.* 8 July 1994.

FM 41-10. *Civil Affairs Operations.* 14 February 2000.

FM 63-4. *Combat Service Support Operations—Theater Army Area Command.* 24 September 1984.

FM 71-3. *The Armored and Mechanized Infantry Brigade.* 8 January 1996.

FM 71-100. *Division Operations.* 28 August 1996.

FM 90-13-1. *Combined Arms Breaching Operations.* 28 February 1991.

FM 90-40. *Multiservice Procedures for the Tactical Employment of Nonlethal Weapons (NLW) {MCRP 3-15.8; NWP 3-07-31; USCG Pub 3-07.31}.* 6 October 1998.

FM 100-1. *The Army.* 14 June 1994.

FM 100-5. *Operations.* 14 June 1993.

FM 100-8. *The Army in Multinational Operations.* 24 November 1997.

FM 100-10. *Combat Service Support.* 3 October 1995.

FM 100-15. *Corps Operations.* 29 October 1996.

FM 100-17-3. *Reception, Staging, Onward Movement, and Integration.* 17 March 1999.

FM 101-5. *Staff Organization and Operations.* 31 May 1997.

FM 101-5-1. *Operational Terms and Graphics (MCRP 5-2A).* 30 September 1997.

JP 3-10. *Joint Doctrine for Rear Area Operations.* 28 May 1996.

USC Title 18. *Crimes and Criminal Procedure.* 23 January 2000.

USC Title 32. *National Guard.* 5 January 1999.

Index

PREFACE

1. Scope

This publication describes multi-Service tactics, techniques, and procedures (MTTP) for consideration and use during the tactical employment of nonlethal weapons (NLW) in support of warfighting personnel conducting training and tactical operations. This publication—

a. Provides an overview of NLW and its relationship to deadly force.

b. Provides NLW system description.

c. Describes the capability requirements of NLW.

d. Discusses fundamental concepts and training requirements involved with NLW.

e. Discusses the tactical employment considerations of NLW.

f. Discusses lessons learned from previous NLW use.

2. Purpose

a. This publication provides a single-source, consolidated reference on the tactical employment of NLW and supporting systems.

b. This publication is not intended to restrict the authority of the commander from organizing the force and executing the mission in a manner deemed appropriate.

3. Application

a. This publication provides commanders and their staffs unclassified guidance for NLW employment and planning. Commanders and staffs can use this publication to aid in the tactical employment of NLW during exercises and contingencies.

b. The United States (U.S.) Army, Marine Corps, Navy, Air Force, and Coast Guard approved this multi-Service publication for use.

4. Implementation Plan

Participating service command offices of primary responsibility will review this publication, validate the information and reference, and incorporate it in service manuals, regulations, and curricula follows.

Army. The Army will incorporate the procedures in this publication in U.S. Army training and doctrinal publications as directed by the Commander, U.S. Army Training and Doctrine Command (TRADOC). Distribution of this publication is in accordance with (IAW) initial distribution number (IDN) 115757.

Marine Corps. The Marine Corps will incorporate the procedures in this publication in U.S. Marine Corps doctrinal and training publications as directed by the Commanding General, U.S. Marine Corps Combat Development Command (MCCDC). Distribution of this publication is IAW Marine Corps Doctrinal Publication System (MCPDS).

Navy. The Navy will incorporate the procedures in this publication in U.S. Navy doctrinal and training publications as directed by the Commander, Navy Warfare Development Command (NWDC). Distribution of this publication is IAW Military Standard Requisition-

MARINE CORPS PCN: 143 000057 00
Air Force Distribution: F

ing and Issue Procedures (MILSTRIP) Desk Guide and Navy Standing Operating Procedure (NAVSOP) Pub 409.

Air Force. The Air Force will validate and incorporate appropriate procedures IAW applicable governing procedures. Distribution is IAW Air Force Instruction (AFI) 33-360.

Coast Guard. The Coast Guard will incorporate the procedures in this publication in U.S. Coast Guard doctrinal and training publications as directed by the Assistant Commandant for Operations, Director of Operations Policy, U.S. Coast Guard (USCG). Distribution of this publication is IAW Coast Guard standing operating procedures.

5. User Information

a. TRADOC, MCCDC, NWDC, Air Force Doctrine Center (AFDC), USCG, and the Air Land Sea Application (ALSA) Center developed this publication with the joint participation of the approving service commands. ALSA will review and update this publication as necessary.

b. This publication reflects current joint and service doctrine, command and control organizations, facilities, personnel, responsibilities, and procedures regarding NLW employment. Changes in service protocol, appropriately reflected in joint and service publications, will likewise be incorporated in revisions of this document.

c. We encourage recommended changes for improving this publication. Key comments to the specific page and paragraph and provide rationale for each recommendation directly to—

*FM 3-22.40 (FM 90-40)
*MCWP 3-15.8
*NTTP 3-07.3.2
AFTTP(I) 3-2.45
*USCG PUB 3-07.31

FM 3-22.40 (FM 90-40)	U.S. Army Training and Doctrine Command Fort Monroe, Virginia
MCWP 3-15.8	U.S. Marine Corps Combat Development Command Quantico, Virginia
NTTP 3-07.3.2	U.S. Navy Warfare Development Command Newport, Rhode Island
AFTTP(I) 3-2.45	Headquarters, Air Force Doctrine Center Maxwell Air Force Base, Alabama
USCG Pub 3-07.31	U.S. Coast Guard Washington, D.C.

15 January 2003

NLW
Tactical Employment of Nonlethal Weapons

TABLE OF CONTENTS

*This publication supersedes FM 90-40, MCRP 3-15.8, NWP 3-07.31, USCG Pub 3-07.31, dated 6 October 1998.

FIGURES

Tables

EXECUTIVE SUMMARY

Tactical Employment of Nonlethal Weapons (NLW)

Today's world events mandate a need to project nonlethal force across all levels of war to enable our warfighters and leaders to effectively deal with a host of traditional as well as nontraditional threats. Now more than ever, the minimal level of tolerance for collateral damage and loss of human life, coupled with the tendency for the typical adversary to exploit the rules of engagement (ROE) to his benefit, necessitates an effective and flexible application of force through nonlethal weapons. Much like a rheostat switch in which power can be dialed up or down as desired, NLW provide tools to allow a commander to employ sufficient force to accomplish an objective without requiring the destruction of an enemy or the habitat. The intent of employing NLW is not to add another step in the progression of escalation with an adversary, but to add another tool to use anywhere along that continuum. This publication provides initial guidance for employing NLW in a tactical environment.

Overview of NLW

Chapter I describes the background, definition, and the concept description of NLW. This chapter discusses the development and employment of NLW, and explains the relationship of NLW to deadly force capabilities. The final portion of the chapter reviews policy per Office of Secretary of Defense instructions on employing NLW.

Purpose, System Description, and Organization

Chapter II describes the purpose of NLW and gives a general description of currently available commercial or government off-the-shelf equipment and munitions. The chapter also addresses organization structure, to include system location/distribution, personnel requirements, and mission requirements/tactical structure.

New Capability Requirements

Chapter III describes the applicability of NLW across the range of military operations. Additionally, this chapter discusses the core capabilities as stated in the Joint Mission Area Analysis (JMAA) as described in the Department of Defense Joint Concept for NLW.

Training With the NLW Capability Set

Chapter IV describes training requirements, topics, methods, locations, and a program of instruction from the Interservice Nonlethal Individual Weapons Instructor Course currently available for NLW options. This chapter also addresses additional training available from civilian law enforcement agencies.

Employment of NLW Capabilities

Chapter V establishes the fundamental understanding of the force continuum needed to employ NLW. It outlines employment prerequisites, mission-planning factors, and personnel requirements needed to employ NLW options. It further delineates NLW tasks according to the JMAA and provides considerations for preparation for unit tactical operations. This chapter also discusses employment objectives for NLW and ROE development. This chapter ends with multiple examples of service-specific crowd control formations used by units equipped with NLW.

Lessons Learned

Chapter VI focuses on NLW lessons learned from recent military operations. The chapter discusses how nonlethal technology has changed the way forces are looking at their adversaries in using NLW. It also covers the use of NLW along the force continuum in conjunction with ROE and the media.

PROGRAM PARTICIPANTS

The following commands and agencies participated in developing this publication:

Joint

U.S. Joint Forces Command, Suffolk, VA
U.S. European Command, J3/J5, APO AE
U.S. Pacific Command, J3832, Camp H. M. Smith, HI
U.S. Central Command, J3, MacDill AFB, FL
U.S. Southern Command, SCJ5-PS, Miami, FL
U.S. Space Command, SPJ5X, Peterson AFB, CO
U.S. Special Operations Command, HQ USSOCOM, SOOP-PTE, MacDill AFB, FL
Joint Nonlethal Weapons Directorate, Quantico, VA
Interservice Nonlethal Individual Weapons Instructor Course, Fort Leonard Wood, MO

Army

Headquarters, Department of the Army, ODCSOPS (DAMO-FDT), Pentagon, Washington, D.C.
U.S. Army Training and Doctrine Command, DCSDCS (ATDO-A), Fort Monroe, VA
U.S. Army Training and Doctrine Command, DCSDEV (ATCD-SN), Fort Monroe, VA
U.S. Army Infantry Center, DCD, Fort Benning, GA
U.S. Army Military Police School, ATTN: DOTD, MANSCEN (ATZT-DT-WM-M), Fort Leonard Wood, MO
USAJFKSWCS, DOTD, Fort Bragg, NC
National Ground Intelligence Center, Charlottesville, VA

Marine Corps

Marine Corps Combat Development Command, Quantico, VA
MARCORSYSCOM, Quantico, VA
USMC Warfighting Lab, Quantico, VA
MCSFT Company, MCSF Battalion, Chesapeake, VA
2d FAST MCSF Battalion, Yorktown, VA
1st MARDIV Military Police Company, HQ Battalion, Camp Pendleton, CA
Marine Corps Detachment Military Police and Chemical School, Fort Leonard Wood, MO
Special Operations Training Group, I Marine Expeditionary Force, Camp Pendleton, CA
Special Operations Training Group, II Marine Expeditionary Force, Camp Lejeune, NC

Navy

Commander, Navy Warfare Development Command (NWDC), Newport, RI
USN Surface Warfare Center, Crane, IN
USN Surface Warfare Center, Dahlgren, VA

Air Force

AFDC/DJ, Langley AFB, VA
Headquarters, ACC/SFOT, Langley AFB, VA
Air Force Security Forces Center/SFOR, Lackland, AFB, TX
Headquarters, USAF/XOF, Washington, D.C.

Coast Guard

Commandant, U.S. Coast Guard, (G-OPL-2), Washington, D.C.

Chapter I

Overview of Nonlethal Weapons

1. Background

a. In today's operational environment, United States (U.S.) forces regularly perform missions that were almost inconceivable a few decades ago. In this new environment, lethal firepower or the threat of its use may no longer be the default solution to all crises or problems. Senior leaders face a new level of public and media sensitivity and scrutiny concerning the proper role of the military as an instrument of national power. Field commanders must understand these sensitivities and attempt to achieve an appropriate military force. Junior leaders must apply the resulting decisions wisely, often in changing situations filled with uncertainty and danger.

b. Throughout history, changes in culture and technology have influenced the character of military force and the manner in which it is employed. These changes are an attempt to maximize military force in the force continuum. In the complex and changing political and social domain of the modern world, nonlethal capabilities may offer the opportunity to increase or maximize the military as an instrument of national power.

c. Nonlethal weapons (NLW) must now be considered an integral part of joint operational capabilities. Joint forces anticipate that NLW technology will advance and provide additional options for force commanders.

2. Definition

a. Department of Defense Directive (DODD) 3000.3, Policy for Nonlethal Weapons, 9 July 1996, defines NLW as "weapons that are explicitly designed and primarily employed so as to incapacitate personnel or materiel while minimizing fatalities, permanent injury to personnel, and undesired damage to property and the environment." Furthermore, "unlike conventional lethal weapons that destroy their targets principally through blast, penetration, and fragmentation, nonlethal weapons employ means other than gross physical destruction to prevent the target from functioning. Nonlethal weapons are intended to have relatively reversible effects on personnel and materiel."

b. The term "nonlethal" does not mean zero mortality or nonpermanent damage; these are goals and not guarantees of these weapons. NLW can add flexibility to combat operations and enhance force protection by providing an environment in which friendly troops can engage threatening targets with reduced risk of noncombatant casualties and collateral damage.

3. Concept Description

Nonlethal capabilities within the context of the service capability sets (see Appendix E) may provide armed forces additional offensive and defensive tools to assist in the incapacitation and destruction of an enemy, capture of important intelligence sources, crowd control, civil disturbances, enforcing mandates, and force protection in accordance with the specified rules of engagement (ROE) for a specific operation. These operations include small-scale contingencies (SSC), military operations other than war (MOOTW), stability and support operations (SASO), and civil disturbances. The U.S. Army's Task Force (TF) FALCON in Kosovo and U.S. armed forces located at Camp X-RAY in Guantanamo Bay, Cuba are superb examples of the use of NLW to prevent escalation within an operation.

4. Background on Development and Employment

a. Military forces have long used nonlethal force to influence behavior of people and nations, defeat adversaries with minimum use of lethal capabilities, and weaken adversaries to defeat them with conventional arms. Examples of nonlethal means include a show of force, physical obstacles, riot control agents (RCA), noise to create or enhance psychological effects, smoke and obscurants to mask operations or defeat homing and guidance mechanisms, and using light to disorient combatants. These nonlethal means will remain relevant in future operations and are a part of an evolutionary process of weapons development. Joint operations in Somalia provided increased awareness of the need to incorporate nonlethal options into the fighting capabilities. Therefore, the Department of Defense (DOD) established the Joint Nonlethal Weapons Program in 1996 to research, develop, and acquire these capabilities.

b. Increased interaction between U.S. forces and civilian populations has become a feature of the contemporary operational landscape that is likely to remain the case in the future. Three factors account for this forecast:

(1) Worldwide patterns of population growth and migration have resulted in an expansion of urban culture within established industrialized nations, as well as many pre-industrial and emerging third world societies (60 percent of the world's population live in built-up areas). The prevalence of urbanization in many crisis-prone regions of the world creates the potential for large, vulnerable groups of noncombatants to be entrapped in volatile confrontations involving the deployment of military forces in military operations in urban terrain (MOUT).

(2) U.S. forces increasingly operate in the challenging environment known as MOOTW; a category that includes such missions as humanitarian assistance, disaster relief, noncombatant evacuations, and the various types of peace operations. These operations commonly involve close and continual interaction between U.S. forces and noncombatant civilians. Some MOOTW scenarios include the presence of paramilitary forces, armed factions, or rogue elements that present a constant, but uncertain threat to joint forces. In these situations, the mission of military forces includes preventive tasks. In other words, U.S. forces enhance mission accomplishment by preventing individuals or factions from carrying out specified undesirable activities, such as rioting, looting, attacking, harassing, or threatening. Often, the adversary blends in with the local population of innocent citizens. At other times, sectors of the local population may rise against U.S. forces and become active participants in acts of violence. Factional alignments, the level of violence, and the threat to mission accomplishment, may change frequently with little or no warning. Under such circumstances, the identity of adversaries is uncertain and the use of deadly force for purposes other than self-defense may be constrained by ROE or the judgment of the on-scene commander.

(3) From a defense-in-depth or situational understanding perspective, the force protection, homeland security, and the antiterrorism missions, require that forces assess the use of NLW as a response to potential asymmetric attacks. Asymmetric threats run the spectrum of adversarial use of innovative or nontraditional strategies, tactics, or technologies, to exploit vulnerabilities and strike a decisive blow while avoiding the strengths of an opponent. The objectives of these strikes are intended to undermine national will. Terrorist attacks such as those directed against the USS Cole, World Trade Center, and Pentagon, highlight the severity of the asymmetric threat and its resultant effects.

5. Relationship of NLW to Deadly Force

a. NLW provide "the lesser means" that could be considered prior to employing deadly force. The services definition of deadly force is as follows:

(1) U.S. Army – "Force that a person uses causing, or that a person knows, or should know, will create a substantial risk of causing death or serious bodily harm." (Army Regulation (AR) 190–14, Military Police Carrying of Firearms and Use of Force for Law Enforcement and Security Duties, 12 March 1993).

(2) USMC – "That force which a person uses with the purpose of causing death or serious bodily harm or which a reasonable and prudent person would consider likely to create substantial risk of causing death or serious bodily harm. It will be used as a last resort when all lesser means have failed." Marine Corps Order (MCO) 5500.6F, Arming of Security and Law Enforcement Personnel and the Use of Force, 20 July 1995.

(3) USN – "Force that a person uses causing, or that a person knows or should know would create a substantial risk of causing death or serious bodily harm." Secretary of The Navy Instruction (SECNAVINST) 5500.29B, Use of Deadly Force and the Carrying of Firearms by Personnel of the Department of the Navy in Conjunction with Law Enforcement, Security Duties and Personal Protection, 29 September 1992.

(4) USAF – "Force that causes, or that you know has a substantial risk of causing death or serious bodily harm." Air Force Instruction (AFI) 31-207, Arming and Use of Force by Air Force Personnel, 1 September 1999.

(5) USCG – "Deadly force is any force that is likely to cause death or other serious physical injury." (COMDTINST M16247.1B, Maritime Law Enforcement Manual, 9 August 2001.)

b. The commitment of military power to resolve crises has traditionally involved the use of deadly force or the implicit or explicit threat of the use of deadly force. Military units are trained, organized, and equipped for this purpose. A force armed only with traditional military weapons normally has two options for enforcing compliance: threating deadly force and applying deadly force. This limitation creates a critical vulnerability that belligerents may quickly discern and use to their advantage.

c. Nonlethal capabilities provide a wider range of options that augment, but do not replace traditional means of deadly force. The option to resort to deadly force must always remain available when the commander believes it is appropriate to the mission. DODD 3000.3, states "the availability of NLW shall not limit a commander's inherent authority and obligation to use all necessary means available and to take all appropriate action in self-defense." The existence of nonlethal capabilities does not represent the potential for "nonlethal war," and unrealistic expectations to that effect must be vigorously avoided. Noncombatant casualties, to include serious injuries and fatalities, are often an unavoidable outcome when employing military power, regardless of NLW availability. Augmenting deadly force with NLW should be fundamental to planning and executing any operation where the commander is contemplating deadly force.

d. The commander must clearly articulate the roles of NLW within the ROE. NLW is an additional means of employing force throughout the force continuum for the particular purpose of reducing the probability of death or serious injury to noncombatants or belligerents. Using deadly force must always remain an inherent right of individuals in instances when they, their fellow servicemen, or personnel in their charge are threatened with death or serious bodily harm. NLW add flexibility to combat operations and enhance force protection by providing the field commander with options to engage threatening targets with limited risk of noncombatant casualties and collateral damage.

e. Commanders and public affairs officers (PAO) must prepare personnel to address media questions and concerns regarding the role of NLW. Operational experience indicates that new capabilities give rise to significant media interest. Personnel participating in interviews or briefings must be prepared to address the role of NLW. It must be clear that the presence of NLW in no way indicates abandoning the option to employ deadly force in appropriate circumstances.

6. Advantages of Employing NLW

NLW provide commanders the flexibility to influence the situation favorably with increased safety to friendly forces and reduced risk of both noncombatant fatalities and collateral damage. Some advantages follow:

a. NLW can be more humane and consistent with the political and social implications implicit in humanitarian and peacekeeping missions.

b. The force that properly employs nonlethal options can gain advantages over those who rely on lethal options alone. Because the degree of provocation required to employ NLW is substantially less than lethal options, the force can take a more proactive position that may diminish the likelihood of having a situation escalate to a point where deadly force is required to resolve a conflict.

c. NLW options can be less likely to provoke others. However, commanders should be cautioned using NLW does not preclude an undesired result.

d. Demonstrated restraint can greatly diminish feelings of anger and remorse when deadly force is required after nonlethal options fail.

e. Using NLW can facilitate post incident stabilization by reducing populace alienation and collateral damage.

f. NLW can reduce the possibility of injury to friendly forces compared to forces without NLW.

7. Policy Per Office of the Secretary of Defense Instructions on Employment of NLW

a. DODD 3000.3 states, "nonlethal weapons, doctrine, and concepts of operation shall be designed to reinforce deterrence and expand the range of options available to commanders. Nonlethal weapons should enhance the capability of U.S. Forces to accomplish the following objectives:

(1) Discourage, delay, or prevent hostile actions.

(2) Limit escalation.

(3) Take military action in situations where use of lethal force is not the preferred option.

(4) Better protect our forces.

(5) Temporarily disable equipment, facilities, and personnel.

(6) Nonlethal weapons should also be designed to help decrease the post-conflict costs of reconstruction."

b. DODD 3000.3 continues by stating, "The availability of nonlethal weapons shall not limit a commander's inherent authority and obligation to use all necessary means available and to take all appropriate action in self-defense. Neither the presence nor the potential effect of nonlethal weapons shall constitute an obligation for their employment or a higher standard for employment of force than provided for by applicable law. In all cases, the United States retains the option for immediate use of lethal weapons, when appropriate, consistent with international law."

c. DODD 3000.3 further states, "Nonlethal weapons shall not be required to have a zero probability of producing fatalities or permanent injuries. However, while complete avoidance of these effects is not guaranteed or expected, when properly employed, nonlethal weapons should significantly reduce them as compared with physically destroying the same target." Nonlethal weapons may be used in conjunction with lethal weapon systems to enhance the latter's effectiveness and efficiency in military operations. This shall apply across the range of military operations to include those situations where overwhelming force is employed."

d. According to paragraph 3.21 of the Interium Guidance issued by the Deputy Secretary of Defense, 30 October 2002, all developmental weapon systems must be submitted for legal review and must receive favorable findings before they can be incorporated into the force. This review includes examination for consistency with applicable international and domestic laws, including the laws of war, various arms control treaties, and protocols. U.S. forces can be assured that any NLW they are issued have passed this critical test. Medical and policy reviews should also be considered prior to fielding any new weapon systems. Commanders and operators should consult their specific ROE for guidance.

e. Some capabilities, although designed to minimize fatalities and serious injuries, may have effects that could actually discourage their use. Therefore, the basic decision to employ military force in defense of national interests is usually a matter of intense public concern. The manner in which that force is exercised is subjected to the same scrutiny.

f. The presence of NLW on the battlefield does not mean they must be used. It also does not mean they must be used prior to the use of lethal force. In all cases, the U.S. retains the option for immediate use of lethal weapons when the on-scene commander deems it appropriate. Additionally, the immediate use of lethal weapons must be consistent with U.S. and international law and the ROE.

Chapter II

PURPOSE, SYSTEM DESCRIPTION, AND ORGANIZATION

1. Purpose

The purpose of NLW is to enhance the ability of military commanders to conduct operations across the spectrum of conflict. Over the past several years, military operations have gravitated towards the lower end of this spectrum and have involved U.S. forces in various missions. These missions include SSC, SASO, MOOTW, and civil disturbances where the ROE are inherently restrictive. NLW are designed to address the special requirements of low-intensity environments. They provide tools that enable the individual, team, or unit, to anticipate and/or respond to provocation with more appropriate means than through the use of deadly force. Essentially, NLW provide U.S. forces with a near-term capability for effectively controlling the nontraditional battlefield, within the constraints levied by ROE, by reducing the risk to friendly forces, mitigating casualties, and minimizing collateral damage.

2. System Description

a. NLW Capability Set. The services have, and are continuing to procure, versatile packages of NLW comprised of commercial off-the-shelf (COTS) and government off-the-shelf (GOTS) equipment and munitions. The government is also exploring advanced technologies for future procurements. In addition to addressing contingency requirements, the services' NLW capability sets provide limited sustainment training ammunition and equipment.

b. NLW Capability Set—Component Categories. Each service determines the contents of its own capability set. Examples of the services current configurations are shown in Appendix E. The NLW capability set components can be divided into four distinct categories: personnel protectors, personnel effectors, mission enhancers, and training devices. The following examples are not all-inclusive.

(1) Personnel Protectors. Personnel protectors include items such as face shields, riot shields, and shin guards that protect the individual from blunt trauma injuries inflicted by thrown objects, clubs, etc.

(2) Personnel Effectors. Personnel effectors include items such as riot batons, malodorants, stingball grenades, pepper sprays, and other kinetic rounds (e.g. sponge grenades) designed to discourage, disorient, or incapacitate individuals or groups.

(3) Mission Enhancers. Mission enhancers include items such as bullhorns, spotlights, caltrops, isotropic radiators (stun/flash), and barriers/entanglers. These items are designed to facilitate target identification and crowd control, and to limit personnel and vehicular movement.

(4) Training Devices. Training devices include items such as training suits, training rounds, training batons, and inert pepper sprays. They are designed to facilitate realistic hands-on scenario training in preparation for operations.

c. System Relationships and Inter/Intraoperability. The NLW capability set is designed to provide an operating force (ranging from squad through battalion-sized units [see Appendix E]) with tactically deployable nonlethal equipment. Fielding of this capability set enhances the operation of systems and equipment already fielded and planned within the joint community. The NLW capability set components are compatible with military equipment that is already fielded or planned for future fielding. The components of the NLW capability set are supported within current military logistic capabilities.

d. Augmentation of Organic Equipment. The operational intent of the capability set is to augment an operating force with the required nonlethal equipment to enhance its force projection capability. The NLW capability set is additional equipment that must be incorporated into lift and load planning.

3. Organization

Organization/organizational structure is not anticipated to change as a result of NLW capability sets being added to the inventory.

a. System Location/Distribution. The United States Army (USA), United States Marine Corps (USMC), and United States Navy (USN) have procured and fielded capability sets to the operating forces, security forces, bases, and stations. All services have current plans to field additional capability sets as required. At the time of this printing, the United States Air Force (USAF) has identified its capability sets to field in the near future. The United States Coast Guard (USCG) also intends to field NLW.

b. NLW Capability Set Composition. Appendix E describes the current and proposed capability sets.

c. Personnel Requirements. To date, no change or increase in military specialty codes have been determined in order to field, operate, and maintain the NLW capability set. Each service determines its own training personnel requirements.

d. Mission Requirements/Tactical Structure.

(1) The mission requirements and tactical structure pertinent to the employment of the NLW will vary depending on the type of contingency and subsequent task organization of the employing unit. Mission requirements, derived from commander's guidance, should be satisfied through the appropriate combination of lethal and nonlethal capabilities.

(2) NLW provide commanders with a means of shaping their area of operations (AO) short of employing deadly force. However, it is imperative that U.S. forces retain the means and ability to employ lethal force in the event that nonlethal force fails to achieve desired outcomes. For example, NLW employment may provoke a lethal response from the adversary. Without the imminent threat of lethal force, nonlethal munitions will prove to be less effective and possibly even counterproductive. More importantly, denying U.S. forces a lethal capability would place them at unacceptable risk. If nonlethal force is employed without the threat of lethal force, the belligerent may opt to exploit a perceived vulnerability. In this case, the use of NLW becomes counterproductive, because it serves only to aggravate the threat and causes the situation to deteriorate.

Chapter III

NLW CAPABILITY REQUIREMENTS

1. Applicability Across the Range of Military Operations

a. U.S. forces conduct military operations across a large spectrum. This spectrum spans from humanitarian assistance and peacekeeping to strategic nuclear exchange. Since the end of the Cold War, the U.S. military has been increasingly involved with operations at the lower end of the spectrum. These operations focus on deterring war, resolving conflict, supporting civil authorities and promoting peace. NLW may be useful across many categories of these operations. The range of military operations includes various levels of combat "intensity," and NLW provide one more force option to deal with the varying intensity. MOOTW may be found across the range of military operations and often include situations where noncombatants are mingled with the adversaries or involved in acts of violence, such as rioting or looting. In such circumstances, NLW provide commanders the flexibility to influence the situation favorably with reduced risk of noncombatant fatalities and collateral damage. NLW also provide a means by which to ascertain the intent of individuals/belligerents. Some characteristics of operations that present the opportunity to use NLW are—

(1) Close contact between friendly forces and noncombatant civilians.

(2) The presence of paramilitary or irregular forces.

(3) Armed civilian gangs, unlawful combatants, or militia.

b. These characteristics present a real but ill-defined threat. Under such circumstances, the commander's judgment or ROE may restrict using deadly force for other than self-defense. NLW may provide U.S. forces flexibility to resolve situations where the use of deadly force is restricted. These weapons provide flexibility by—

(1) Allowing U.S. forces to apply measured military force.

(2) Reducing risk of serious noncombatant causalities.

(3) Accomplishing the mission.

(4) Providing force protection.

(5) Retaining the initiative.

(6) Allowing U.S. forces to remain relevant and capable in increasingly complex and chaotic environments.

c. The need to reduce the risks of serious injury to personnel is not limited to crowd control scenarios. In MOUT, some of the local civilian populace may remain in an urban area in the midst of battle. The traditional solution to such challenges has been restricting ROE. Nonlethal capabilities may provide commanders the flexibility to adapt a fluid approach to MOUT, which allows their subordinates the freedom of action to employ appropriate levels of measured military force to accomplish their mission.

d. Another example of NLW applications in MOOTW is in peace enforcement. In these operations, NLW are used in an area denial role to separate belligerents without resorting to deadly force. They are also used to support humanitarian relief organizations in protecting food and medical distribution sites. The presence of NLW will not limit a commander's inherent authority and obligation to use all necessary means available and to take appropriate action in self-defense.

2. Core Capabilities

Core capabilities are those fundamental competencies that enable U.S. forces to achieve desired operational objectives. NLW provide a flexible means to protect friendly forces, high-value assets, and influence the actions of potential adversaries and/or noncombatants. These goals can be achieved without resorting to lethal force and in a manner that may minimize collateral damage. The core capabilities associated with nonlethal effects fall into three major categories as stated within the Joint Mission Area Analysis (JMAA) for Nonlethal Weapons (approved December 2000 by the Joint Requirements Oversight Council): counterpersonnel, countermateriel, and countercapability.

a. Counterpersonnel Capabilities. The DOD Joint Concept for Nonlethal Weapons concept paper, 5 January 1998, describes required nonlethal counterpersonnel capabilities for U.S. forces as follows:

(1) "The application of military force to accomplish a mission with reduced risk of fatalities or serious casualties among noncombatants—or even, in some instances, among enemy forces." There are several specific nonlethal counterpersonnel capabilities to be considered. These include, but are not limited to, the means to influence the behavior and activities of a potentially hostile crowd and the capability to bring a mob engaged in a riot under control. While there are many similarities in these two capabilities, each involves unique challenges, which may require radically different solutions.

(2) "Capability to incapacitate personnel." For the purposes of this document, "incapacitation" is achieved when the weapon's effects render personnel unable to physically or mentally (real or perceived) act in a hostile or threatening manner. In keeping with the guiding principles for NLW, this incapacitation should be reversible through the passage of time. The focus will be on capabilities that affect both individuals and groups of people.

(3) "Capability to deny personnel access to an area (land, sea, or air)." NLW can provide new possibilities for barrier planning in any type of military operation, including systems that produce physical or mental discomfort to those who enter the denied area.

(4) "Capability to clear facilities and structures of personnel." The Joint Concept for NLW states this capability "will facilitate MOUT by reducing the risks of noncombatant casualties and collateral damage while simultaneously minimizing the advantages accruing to an enemy defending a built up area." This capability is applicable to all clearing missions that may be conducted (e.g., Maritime Interception Operations [MIO]) and could assist in providing enhanced force protection.

(5) "Capability to seize personnel, which could include some combination of technologies inherent in other core capabilities, for example, incapacitation and entangling devices (which may aid in seizing personnel)." This capability is intended to augment lethal means used to capture specified individuals, such as enemy combatants or persons who are inciting a mob or crowd to violence.

b. Countermateriel Capabilities. The Joint Concept for Nonlethal Weapons describes countermateriel capabilities as follows:

(1) Rendering equipment and facilities unusable without complete destruction by attacking only weapons of war and supporting infrastructure while minimizing risk to noncombatants. This could enhance operations by defusing potentially volatile situations where conventional military means might otherwise prove counterproductive.

(2) Disabling or denying the use of vehicles, vessels, and aircraft entry into targeted areas or access within an AO. This capability may include physical barriers that impede or stop movement. Alternative systems may alter combustion properties of fuels

and viscosity of lubricants, embrittle or decay materials, attack tires, gaskets, or hoses, or act as adhesives to fuse metal parts. The requirement also includes the capability to facilitate boarding and clearing operations.

c. Counter Capability Capabilities. For U.S. forces counter capability capabilities are described within the JMAA for NLW as follows:

(1) "Disable or neutralize facilities and systems." This includes a nonlethal capability for disabling/neutralizing electrical generating facilities, command, control, communications, computers, intelligence, surveillance, and reconnaissance (C4ISR) systems, Integrated Air Defense Systems (IADS), weapons systems, optical sensors, electrical sensors, and navigation capabilities with such controlled effects as to allow for selective, precise engagement. Other needs include deceiving reconnaissance with illusions of barriers, obstacles, and/or forces, and concealing/covering friendly activities or movements, particularly within complex urban terrain.

(2) "Deny the use of weapons of mass destruction (WMD)." Mission needs include rendering a WMD inoperative; containing the potential release of deadly agents/contaminants; and preventing or neutralizing the production, storage, deployment (transport), employment, and delivery of WMD. Nonlethal capabilities in combating WMD should be designed to permit engagement within a populated or sensitive terrain, prior to the initiation of armed conflict.

3. Limitations

The limitations of NLW are effectiveness, perception, and legality.

a. The effectiveness of NLW is dependent on factors such as motivation, age, environmental and human factors, range of engagement, and numerous others. NLW can be lethal if improperly applied. Furthermore, countermeasures for thwarting virtually all nonlethal options are usually apparent, quickly learned, and readily available. Because they are not intended to kill, nonlethal options teach adversaries what to avoid in the future. Small unit commanders must prepare to keep one step ahead of belligerents.

b. The perceptual limitation is that personnel may misunderstand the appropriate applications of NLW across the range of military operations. The incorrect perception that NLW will allow wars and MOOTW to be prosecuted without casualties may lead to conflicting expectations between political and military leaders. These conflicts can result in U.S. forces being vulnerable to misplaced or inconsistent ROE. All leaders, political and military, involved in planning and executing military missions must understand there are no "nonlethal operations."

c. Some proposed NLW might be forbidden by law or policy. Accordingly, it is essential that all NLW developments be evaluated by appropriate authorities to ensure they comply with the law of war, U.S. law, and U.S. treaty obligations. Use of NLW containing chemical agents, for instance, must be evaluated in the context of the Chemical Weapons Convention. Also, using RCA in an armed conflict requires Presidential approval.

Chapter IV

TRAINING

1. Fundamental Concepts and Training Requirements

The successful accomplishment of any operation in which nonlethal measures are employed requires extensive preparation, of which individual, unit, medical support, and unit training are vital parts. Training should be designed to give individuals an understanding of the entire subject area and enable them to function efficiently as members of a unit. This training must be intensive and realistic. Training with the NLW capability set, and other NLW, is critical to employ these weapons with confidence. The training flow should be—train leaders, train instructors, train unit, exercise. Qualified instructors should conduct initial unit training. Unit sustainment training should be conducted per individual service and unit requirements.

2. Training of Instructors

a. Verbal communication skills, open-hand control techniques, RCA, effects of impact weapons, and civil disturbance tactics, techniques, and procedures (TTP) are areas trained to standard at the Interservice Nonlethal Individual Weapons Instructor Course (INIWIC) held at the Marine Corps Detachment, Fort Leonard Wood, Missouri. INIWIC is an Interservice Training Review Organization (ITRO) course, and quotas are available to any service through normal training reservation channels. A course outline is provided in Appendix C. This is the only formal DOD NLW instructor-training course. The training requirements outlined in the INIWIC represent the minimum training standards to include in developing NLW and tactics instructors. Additionally, the United States Army Military Police School (USAMPS) has developed a training support package that includes Army-specific TTP for small units (company level and below) in using NLW capabilities. The reference for this is the U.S. Army's website at: www.wood.army.mil.

b. The INIWIC is a "train-the-trainer" program to provide commanders NLW instructors within their command. Furthermore, the military services are not precluded from developing their own training based upon the INIWIC curriculum. The commander can then tailor the NLW instruction to fit mission needs. Establishing the service-level INIWIC serves as an assurance measure, giving the commander a quantifiable means of certifying his nonlethal instructors.

3. Nonlethal Training Topics for Deploying Units

Each service and unit will have unique training requirements based on mission type, unit size, deployment area, etc. As a minimum, the training instructor should cover the following topics in any training for units deploying with nonlethal capabilities. They are part of the INIWIC curriculum.

NOTE: U.S. Army training standards are contained in Army Regulation (AR) 350-38, Training Devices Policies and Management, 15 October 1993; USMC training standards for NLW are contained in MCO 1510.112, Individual Training Standards for Nonlethal Weapons, 27 May 1998.

a. Force Continuum. This is an introduction to the federal force continuum model as outlined by MCO 5500.6F, Arming of Law Enforcement and Security Personnel and the Use of Deadly Force; AR 190-14; AFI 31-207, Arming and Use of Force by Air Force Personnel (1 September 1999) (or other service orders). Identifying the proper levels of force, understanding levels of resistance, and how nonlethal technologies affect the force continuum are the minimum topics the instructor should cover.

b. Crowd Dynamics/Crowd Control. This is an introduction to the differences between crowds, mobs, and riots. This course teaches the student basic crowd control techniques that will be easily applied to various situations. The student should be familiarized with tactics and techniques found in field manual (FM) 19-15, MCWP 3-11.1, Rifle Company, TC 90-1, and Air Force manual (AFMAN) 31-201V6, Civil Disturbance (1 April 2002), but will also consider nontraditional and small unit application.

c. Communication Skills. This is an introduction to the stages of conflict management, verbal aggression, nonverbal communication, physical aggression, physiological diversions, and proper mental conditioning. The first NLW skill an individual must have is the ability to communicate. General A. M. Gray, former Commandant of the Marine Corps, once said, "A warrior's most formidable weapon is his mind." The results of an individual's mental processes are evident through actions and words, and the words and communication skills used in dealing with people are truly the first line of nonlethal options.

d. Oleoresin Capsicum (OC) Aerosol and other RCA Training. This is an introduction to the uses of OC/pepper spray and other RCA. The student should have an appreciation for decontamination requirements, storage, and shelf life of OC/RCA. The instructor will also teach legal and tactical considerations. INIWIC certified instructors should teach existing training programs that include live agent training.

e. Open-Hand Control. This is an introduction to pressure point control techniques, unarmed self-defense measures, weapon retention techniques, and other submission/restraint/search techniques. The instructor should cover handcuffing/flex-cuffing for contact team members.

NOTE: USMC plans to replace the current open-hand defensive tactics training with the USMC Martial Arts Program.

---

f. Impact Weapons. This is an introduction to using the rigid straight baton, collapsible straight batons, side handle batons, and/or riot control batons.

g. ROE/Law of Armed Conflict (LOAC). This is an introduction to the ROE/LOAC and their relationship. (Refer to DODD 5100.77, AR 27-1, and NWP 1-14M).

h. Nonlethal Munitions—Employment and Tactics. This subcourse is an introduction to the capabilities and employment of available NLW. NLW provide additional unit force protection capabilities. Introduction to tactics is outlined in Chapter 5. Appendix D is a sample unit training schedule covering the topics listed above. U.S. Army Training and Doctrine Command (TRADOC) has developed training support packages that will include company level and below TTP.

4. Leader Training

a. An understanding of the topics covered in the INIWIC curriculum is essential. However, commanders/leaders should also possess an in-depth understanding of the ROE, their implementation, as well as the politics, culture, and demographics of the area to which they are deploying.

b. Commanders should involve their medical personnel, public affairs (PA) representatives, civil affairs (CA), psychological operations (PSYOP), staff judge advocate

(SJA), engineers, explosive ordnance disposal (EOD), and military police (MP) in training exercises when possible. Involvement by these special staff sections will greatly enhance the quality and realism of the training. Medical personnel will better understand OC decontamination and the possible injuries they may encounter as a result of NLW employment. Because of the international attention that NLW receive, all personnel will benefit from media training by PA personnel. CA and PSYOP personnel can provide area-specific cultural, ethnic, and political briefs to enhance awareness of the challenges within the AO. SJA personnel can clarify complex ROE questions and assist in training on ROE-related issues. Additionally, supply and logistic staff members also require training in NLW maintenance and environmental issues unique to the equipment and munitions.

5. Unit Training

The INIWIC is primarily designed to produce instructors who will train individuals in using and employing NLW. Collective unit tactical training is the unit's responsibility; however, a NLW sample schedule based on INIWIC training is provided in Appendix D. Units should conduct collective training before deploying with NLW. Unit standard operating procedures (SOP) and tactics may require adjustment to enable the employment of NLW and new equipment for the success of specific missions. All collective training should include nonlethal options and decision making. Whenever NLW are employed during training, the instructor must also conduct lethal capability training. Troops must understand the integration of nonlethal and lethal capabilities and always be prepared to protect themselves, as required, with lethal force. It is impossible to accurately predict the response of an individual or crowd.

6. Training Disclaimer

Initial unit training should not be conducted without qualified instructors. Non-type classified and COTS nonlethal munitions cannot be fired without Army Materiel Command (AMC), Army Training Support Center (ATSC), Crane Naval Weapons Center, or Air Force Munitions Center type classifying the ammunition. The instructor should conduct unit sustainment training in accordance with (IAW) service guidelines.

7. Additional Training

a. Additional training is left to the discretion of the unit but should be encouraged in order to add credibility to the training package. Some recommended training certifications for instructors include, but are not limited to—

(1) Management of Aggressive Behavior (MOAB)
R.E.B. Training International, Inc.
PO Box 845, Stoddard, NH 03464

(2) Oleoresin Capsicum Aerosol Training (OCAT)
R.E.B. Training International, Inc.
PO Box 845, Stoddard, NH 03464

(3) Monadnock Defensive Tactics System (MDTS)
Training calendar available at the Monadnock website:
http://www.batons.com/traincal.html

b. These courses add viable credentials to the instructors as perceived by civilian and military alike. In addition, the instructor can issue certification cards for the area which an individual has been certified. Certifications in various areas offers additional protection in the event of a lawsuit or claim against the government by ensuring the individuals that have been trained within your unit are completely capable.

8. Summary

NLW and tactics training are not meant to be stand-alone training—units must anticipate the requirement for NLW in their missions and train accordingly. NLW provide commanders with a graduated response in the force continuum.

Chapter V

EMPLOYMENT

1. Background

a. A military unit deployed to an area will normally conduct operations IAW current doctrine. The environment may contain both noncombatants and combatants. Nonlethal capabilities provide the military force with tools to engage adversaries and minimize collateral damage. Fundamental to employing NLW is a thorough understanding of the force continuum. As stated before, nonlethal capabilities provide a wider range of options that augment traditional means of deadly force but do not replace them. Nonlethal capabilities allow a commander the flexibility to increase and decrease the amount of force applied to accomplish a mission. The force continuum is generally continuous and seamless, yet a careful examination reveals five broad categories:

(1) Threats. The force continuum begins with threats. These can be either implied or expressed. An implied threat may be manifested through presence by forming into riot-control formations, fixing bayonets, or donning protective masks. An expressed threat occurs when a commander makes known the consequences of defiance.

(2) Denial options. The equipment used in denial options are usually placed in highly visible areas and often marked with placards. Examples include concertina, caltrops, barbed wire, or other obstacles.

(3) Munitions that cause physical discomfort/incapacitate. Munitions that cause physical discomfort, but fall short of inflicting trauma, include flash bangs, RCA—tear gas (O-Chlorobenzyl-Malononitrile [CS]), and OC. Although the discomfort or injury may be substantially less than impact munitions, employing these munitions requires an appropriate use of force assessment. Factors such as training, discipline, prejudices, emotions, and judgment all play a part in their application.

(4) Munitions that inflict blunt trauma. Munitions that inflict blunt trauma will inflict painful or debilitating injuries. These munitions constitute the upper end of nonlethal options within the force continuum. Examples include foam batons, stingball grenades, beanbags, sponge projectiles, and rubber ball rounds.

(5) Lethal munitions. Lethal weapons are at the highest level of force continuum. Lethal options should always be regarded as part of the force continuum and not as a separate option altogether. ROE, along with judgment and experience, helps on-scene commanders decide when using lethal force is authorized and appropriate.

b. Nonlethal capabilities include using existing systems that were not designed specifically as NLW. Nonlethal capabilities can minimize fatalities, permanent injury to personnel, and undesired damage to property, facilities, and the environment. Examples include PSYOP, CA, military working dogs, and PA. However, this chapter limits discussion primarily to using weapons and components explicitly designed as part of a NLW capability set. This focus is not intended to limit the commander's option to use whatever means are available to accomplish the mission without unnecessary lethality.

2. Mission Planning Factors

Incorporating NLW into a plan does not change the fundamental principles of mission planning. Commanders should use NLW as an operational enhancement. The first concern of a commander is successfully completing the mission. ROE are a critical consideration in the commander's planning and guide him in determining the appropriate use of nonlethal and lethal force. The availability of NLW does not relieve a force from its obligation to defend itself. Therefore, it must be equipped and trained with both lethal and nonlethal means.

3. NLW Tasks

a. NLW tasks were derived from the combatant commanders and service mission needs statements and are codified in the JMAA for Nonlethal Weapons. These tasks are grouped under three core capabilities: counterpersonnel, countermateriel, and countercapability.

(1) Counterpersonnel:

(a) Control crowds.

(b) Incapacitate personnel.

(c) Deny an area to personnel.

(d) Clear facilities/structures of personnel.

(2) Countermateriel:

(a) Deny an area to vehicles, vessels, and aircraft.

(b) Disable/neutralize vehicles, vessels, aircraft, and equipment.

(3) Countercapability:

(a) Disable or neutralize facilities and systems.

(b) Deny use of WMD.

b. Although not all encompassing, the following list of employment considerations is designed to assist commanders and staffs in formulating plans for use of NLW:

(1) General considerations:

(a) Understand and define ROE.

(b) Commander develops identification of friend or foe (IFF) procedures.

(c) Distribute NLW munitions to troops in advance.

(d) NLW options require lethal overwatch fire.

(e) Ensure area is well lighted to include perimeter.

(f) Employ night vision devices outside the perimeter to identify personnel.

(g) Utilize appropriate media to inform civilians to stay away from denied areas.

(h) Enforce positive access control into perimeter (personnel/vehicles).

(i) Ensure posted signs are easily understood (symbols/pictures). If interpreter support available, post warning signs in native language.

(j) Use entanglements such as caltrops, tanglefoot, barbed wire, and concertina/razor tape as these disrupt/slow intrusion attempts.

(k) Employ sensors, if available.

(l) Employ aerial or naval pickets.

(2) Counterpersonnel considerations:

(a) Employing kinetic rounds at less than 15 feet/5 meters may result in fatal outcomes.

(b) Target area at 15 to 40 feet should be center mass for blunt munitions.

(c) Head shots may be lethal.

(d) Kinetic rounds will not be skip fired.

(e) Kinetic rounds are effective against selected individual targets and can be used to disperse individuals.

(f) Considering differing rates of fires; the M203 and shotgun are complementary in a nonlethal role, depending upon the munitions used and their effective ranges.

(g) Employ riot batons only after proper training.

(h) Use RCA to disrupt/disperse crowds.

(i) Use RCA markers (RCA with a nontoxic, water-soluble marking dye) to mark agitators for future identification and apprehension.

(j) RCA greatly reduce visibility of control force.

(k) Closely control the employment of RCA.

(l) Disperse RCA to face and eyes for full effect.

(m) Employ optics with designated marksmen to identify mob leaders.

(n) Searchlights dazzle/disorient individuals (night only).

(o) Use bullhorn for communicating with crowd and to control formations.

(p) Employ recovery/apprehension/snatch teams to apprehend agitators.

(q) Use flex cuffs for agitators.

(r) Prepare disposition plan for cleared personnel.

(s) Isolate area to prevent reinforcements.

(t) Consider future friendly use of cleared facility.

(u) Prepare security plan for cleared facility.

(v) Channel passive traffic away from formations.

(w) Maintain rapid reaction force/quick reaction force (RRF/QRF).

(3) Countermateriel/countercapability considerations:

(a) Use marking agents to mark a vessel for later interception.

(b) Consider materiel handling equipment required to evacuate vessel/airframe.

c. Selected NLW tasks are further addressed in Tables V-1 through V-7. The tables are not intended to be all-inclusive but merely a stimulus to planning. The equipment listed is taken from the currently available systems designed for nonlethal purposes. The user should not limit his nonlethal options to this specific equipment and should be aware that there are many existing and emerging systems and potential uses of current technology and common objects to achieve the desired mission end state.

Table V-1. Crowd Control

Tasks	*NLW Planning Factors*	*Equipment*	*Employment Considerations*	*Notes*
Crowd control	Administration/ Personnel • SJA / claims planning • Chaplain affairs • Casualty notification • Detainee disposition plan • PAO/media planning • Information dissemination	Support Equipment • Riot face shield • Full length riot shield • Expandable riot baton • Optics • Portable bullhorn • Vest	• Distribute to troops in advance • Employ riot baton only after proper training • Employ optics with designated marksman (DM) to identify (ID) mob leaders • Use bullhorn for communicating with crowd and to control formations	• Strictly define ROE • NLW options require lethal cover fire
Crowd control	Intelligence • Intelligence preparation of the battlespace (IPB) • Background on country and culture • Route/map recon • Sensor emplacement for advance notification • Information on Mob leaders • Patrols and listening post (LP)/observation post (OP) • Determine reason for the crowd	Kinetic Rounds • 12-gauge (ga) bean bag round • 12-gauge rubber bullet • 12-gauge launching cartridge • 40-mm rubber baton • 40-mm wooden baton **NOTE:** USMC no longer uses wooden baton. • 40-mm stinger grenade • Stun grenade • Flash bang	• Do not employ kinetic rounds at less than 15 feet due to possible fatal outcome • Target area at 15 to 40 feet should be center mass • Head shots are not acceptable • Kinetic rounds will not be skip fired • Kinetic rounds effective against selected targets and to disperse individuals • M203 and shotgun are nonlethal complementary; M203 has low rate of fire; area target. Shotgun has high rate of fire; point target	

Table V-1. Crowd Control

Tasks	***NLW Planning Factors***	***Equipment***	***Employment Considerations***	***Notes***
Crowd control	Operations • Tactical Integration of NLW and lethal fires • Fire Support Element (FSE) plans to include RCA • Allocation of NLW munitions • Integration of NLW fires and maneuver • Engineer barrier planning • Understand mission (deny access/disperse/ monitor)	RCA • OC Dispenser • Team OC dispenser • High volume dispenser	• Disrupt/disperse crowds • Disperse RCA to face and eyes for full effect • Employ snatch teams to apprehend agitators	
Crowd control	Logistics • Distribution of NLW supplies • Security of rear area facilities • Resupply of NLW munitions • Procurement of medical supplies • Hospital planning • Medic planning for tactical units	Riot Control • Xenon Searchlight • Flex cuffs • RCA markers	• Searchlights dazzle/disorient individuals (night only) • Flex cuffs used for agitators • RCA markers are used to mark agitators for future ID	

Table V-2. Incapacitate Personnel

Task	*NLW Planning Factors*	*Equipment*	*Employment*	*Notes*
Incapacitate personnel	Administration/Personnel • SJA/claims planning • Media plan	Support Equipment • Riot face shield • Full length riot shield • Expandable riot baton • Rifleman's combat optic • Vest	• Distribute to troops in advance • Do not employ riot baton to the head • Employ optic with DM to ID mob leaders and cover with lethal fire	• Commander needs to define incapacitation mission • Take a balanced approach to public dissemination of NLW so counter measures cannot be rapidly developed • Cover NLW options by lethal force
Incapacitate personnel	Intelligence • ID types of insurgents and demonstrators • ID causes and factions • ID mob ringleaders	Kinetic Rounds • 12-gauge bean bag round • 12-gauge rubber bullet • 12-gauge launching cartridge • 40-mm rubber baton • 40-mm wooden baton **NOTE:** USMC no longer uses wooden baton. • 40-mm stinger grenade • Stun grenade • Flash bang • M203 and shotguns are nonlethal complementary. M203 has a low rate of fire, area target. Shotgun has a high rate of fire, point target	• Do not employ kinetic rounds at less than 15 feet due to possible fatal outcome • Target area at 15-40 feet should be center mass • Head shots are not acceptable • Kinetic rounds will not be skip fired • Kinetic rounds effective against selected targets and to disperse individuals • M203 and shotguns are nonlethal complementary. M203 has a low rate of fire, area target. Shotgun has a high rate of fire, point target	

Table V-2. Incapacitate Personnel

Task	*NLW Planning Factors*	*Equipment*	*Employment*	*Notes*
Incapacitate personnel	Operations • Prepare and confirm ROE for NLW • Integration of NLW and deadly force • Allocate NLW and designate unit use • Integrate NLW fires and maneuver	RCA • OC dispenser • Team OC dispenser • High-volume • OC dispenser	• Disrupt/disperse crowds • Disperse RCA to face and eyes for full effect	
Incapacitate personnel	Logistics • Distribute NLW munitions • ID special storage requirements • Resupply NLW munitions	Riot Control • Xenon • searchlight • Flex cuffs • Chemical markers	• Searchlights dazzle/disorient individuals (night only) • Use flex cuffs for agitators • Use RCA markers to mark agitators for incapacitation	
Incapacitate personnel	Civil Affairs • Collect civilian intelligence • Care for injured personnel • Detain civilian personnel			

Table V-3. Deny Area to Personnel

Task	*NLW Planning Factors*	*Equipment*	*Employment*	*Notes*
Deny area to personnel	Administration/ Personnel • SJA/claims planning • Media plan	Support Equipment • Area lights • Area sensors • Night vision • Access control • Warning signs written in host nation language	• Ensure area is well lighted to include perimeter • Employ night vision devices outside the perimeter • Positively ID accessed personnel • Employ warning signs in native language	• Utilize appropriate media to warn civilians away from denied area

Table V-3. Deny Area to Personnel

Task	*NLW Planning Factors*	*Equipment*	*Employment*	*Notes*
Deny area to personnel	Intelligence • Terrain and map study • Threat ID • Sensor emplacement around area • IPB			
Deny area to personnel	Operations • Prepare and confirm ROE • Integrate NLW and deadly force • Allocate NLW and designate unit use • Determine scope of mission • Determine access procedures • Use of area (friendly or total denial)			
Deny area to personnel	Logistics • Distribute NLW munitions • ID special storage requirements • Resupply NLW munitions • Security of rear area facilities.			
Deny area to personnel	Civil Affairs • Collect civilian intelligence	Entanglements • Caltrops • Tanglefoot • Barbed wire • Foams • Concertina/ razor tape • Stinger spike system	• Disrupt/slow intrusion attempts • Maintain suitable reaction force • Channel passive traffic away • Use foam to cover caltrops	

Table V-4. Deny Area to Vehicles

Task	*NLW Planning Factors*	*Equipment*	*Employment*	*Notes*
Deny area to vehicles	Administration/ Personnel • SJA/claims planning • Media plan	Support Equipment • Area lights • Area sensors • Night vision • Access control • Warning signs written in host nation language	• Ensure area is well lighted to include perimeter • Employ night vision devices outside the perimeter • Positively ID accessed vehicles	• Commander needs to define denial mission • Utilize appropriate media to warn civilians away from denied area
Deny area to vehicles	Intelligence • Terrain and map study • Threat ID • Sensor emplacement around area • IPB • Traffic analysis			
Deny area to vehicles	Operations • Prepare and confirm ROE • Allocate NLW and designate unit use • Determine scope of mission • Determine access procedures • Determine MP/ security forces requirements • Use of area (friendly or total denial)			• Commanders must consider flying operations when employing RCA. Area(s) should be reusable in short order.
Deny area to vehicles	Logistics • Distribute NLW supplies • Be prepared to evacuate stalled vehicles			
Deny area to vehicles	Civil Affairs • Collection of civilian intelligence	Entanglements • Caltrops • Tanglefoot • Barriers • Organic equipment • Barbed wire • Foams	• Disrupt/slow intrusion attempts • Maintain suitable reaction force • Channel passive traffic away • Use foam to cover caltrops	• Commanders must consider flying operations when employing NLW. NLW should not hinder flying operations.

Table V-5. Deny/Neutralize Vehicles, Aircraft, Vessels, and Facilities

Task	***NLW Planning Factors***	***Equipment***	***Employment***	***Notes***
Deny/ neutralize vehicles, aircraft, vessels, facilities	Administration/Personnel • SJA/claims planning • Media plan	Support Equipment • Sensors to track approaching vehicles, aircraft, and vessels • Night vision equipment • Materiel handling equipment • Portable bullhorn	• Employ ground and water sensors, if required • Employ aerial or naval pickets • Use materiel handling equipment to evacuate vessel	• Commander needs to define neutralization mission • Element of surprise remains a critical factor in mission accomplishment • Commander needs to develop IFF procedures
Deny/ neutralize vehicles, aircraft, vessels, facilities	Intelligence • Sensor emplacement if required • Threat ID • IPB • Analyze target area			
Deny/neutralize vehicles, aircraft, vessels, facilities	Operations • Prepare and confirm ROE • Integrate NLW and deadly force • Allocate NLW and designate unit use • Determine scope of mission • Determine access procedures • Determine boarding procedures (vessels and aircraft) • Determine apprehension procedures • Determine decontamination (DECON) procedures	RCA • Marking agent	• Use marking agents to mark a vessel for later interception	

Table V-5. Deny/Neutralize Vehicles, Aircraft, Vessels, and Facilities

Task	*NLW Planning Factors*	*Equipment*	*Employment*	*Notes*
Deny / neutralize vehicles, aircraft, vessels, facilities	Logistics • Distribute NLW munitions • ID special storage requirements • Resupply NLW munitions • Determine vessel/aircraft disposition	Riot Control • Xenon searchlight	• Use RCA to incapacitate crew members • Searchlights can disorient personnel or illuminate target	• Commanders must consider flying operations when employing RCA • Area(s) should be reusable in short order.
Deny / neutralize vehicles, aircraft, vessels, facilities	Civil Affairs • Collect civilian intelligence • Handle displaced civilian personnel	Entanglements • Caltrops • Foams • Speed bumps • Nets • PVAB	• Use caltrops to disable vehicle • Use combination of foam and caltrops to preserve the element of surprise • Channel passive traffic away • Use nets to foul props	
Deny/neutralize vehicles, aircraft, vessels, facilities	Medical • Treating injured civilian personnel			

Table V-6. Seize Personnel

Task	*NLW Planning Factors*	*Equipment*	*Employment*	*Notes*
Seize personnel	Administration/ Personnel • SJA/claims planning • PAO media plan	Support Equipment • Optics	• Employ optics with DM to ID personnel to be seized • Cover personnel employing nonlethal force	• Commander needs to define seizure mission and strictly enforce ROE • Element of surprise is critical for mission accomplishment • Speed is critical • Seizing personnel can be surgical with special operations personnel or conducted within a crowd to seize an agitator
Seize personnel	Operations • Prepare and confirm ROE • Integrate NLW and deadly force • Allocate NLW and designate unit use • Integrate NLW fires and maneuver • Breaching and assault planning • Determine holding area/processing center	RCA • OC dispenser • CS	• Use RCA to incapacitate personnel • Spray into eyes and face for full effect	
Seize personnel	Logistics • Distribute NLW munitions • ID special storage requirements • Resupply NLW munitions • Plan transportation	Riot Control • Xenon searchlight • Flex cuffs • RCA markers	• Searchlights dazzle/disorient individuals • Use flex cuffs for apprehension • Use RCA markers to mark agitators for seizure	

Table V-6. Seize Personnel

Task	*NLW Planning Factors*	*Equipment*	*Employment*	*Notes*
Seize personnel	Civil Affairs • Collect civilian intelligence • Care for injured personnel • Detain civilian personnel			
Seize personnel	Medical • Medical evacuation (MEDEVAC) planning			

Table V-7. Clear Facilities of Personnel

Tasks	*NLW Planning Factors*	*Equipment*	*Employment Considerations*	*Notes*
Clear facilities of personnel	Administration/ Personnel • SJA/claims planning • Chaplain affairs • Casualty notification • PAO media plan	Support Equipment • Riot face shield • Full-length riot shield • Expandable riot baton • Optics • Portable bullhorn	• Distribute to troops in advance • Do not employ riot baton to the head • Employ optic with DM to ID occupant leaders and provide lethal cover fire • Use bullhorn used for communicating with occupants	• Surprise is a critical element for mission accomplishment • Commander must strictly define ROE • Due to close quarter tactics, the force must be instantly ready to utilize deadly force if required

Table V-7. Clear Facilities of Personnel

Tasks	*NLW Planning Factors*	*Equipment*	*Employment Considerations*	*Notes*
Clear facilities of personnel	<u>Intelligence</u> • IPB • Background on facility, access, power plant • Route/map recon • Background info on building occupants • Determine any outside support for building occupants	<u>Kinetic Rounds</u> • 12-gauge bean bag round • 12-gauge rubber bullet • 12-gauge launching cartridge • 40-mm rubber baton • 40-mm wooden baton **NOTE:** USMC no longer uses wooden baton. • 40-mm stinger Stun grenade • Flash bang	• Do not employ kinetic rounds at less than 15 feet due to possible fatal outcome • Targeted personnel at 15 to 40 feet should be center mass • Head shots are not acceptable • Kinetic rounds will not be skip fired • Kinetic rounds effective against selected targets and to disperse individuals • Stun grenade effects can be amplified in buildings • Visibility will be reduced once kinetic rounds are employed • M203 and shotguns are nonlethal complementary. M203 has a low rate of fire, area target. Shotgun has a high rate of fire, point target	• Using RCA greatly reduces visibility • Use of RCA is restricted to certain circumstances; verify that proper authority has been given before use

Table V-7. Clear Facilities of Personnel

Tasks	*NLW Planning Factors*	*Equipment*	*Employment Considerations*	*Notes*
Clear facilities of personnel	Operations • Tactically integrate NLW and lethal fires • FSE plans to include RCA • Allocate NLW munitions • Integrate NLW fires and maneuver • Determine assault and breaching plan • Plan for isolation of facility	RCA • OC dispenser • Team OC dispenser • High-volume dispenser	• Disrupt/disperse crowds • Apply RCA to face and eyes for full effect	• Isolate area to prevent reinforcements • Prepare disposition plan for cleared personnel • Consider use of facility • Prepare security plan for cleared facility
Clear facilities of personnel	Logistics • Distribute NLW supplies • Resupply NLW munitions • Plan for transportation of displaced civilians • Plan facility maintenance or repair	Riot Control • Xenon searchlight • Flex cuffs • RCA markers • Flash bangs	• Searchlights dazzle/disorient individuals • Use flex cuffs for agitators • Use RCA markers to mark agitators for future ID • Flash bangs to dazzle/disorient individuals	
Clear facilities of personnel	Civil Affairs • Collect civilian intelligence • Care for injured civilians • Coordinate plans for controlling civilians			
Clear facilities of personnel	Medical • Hospital planning • Medic planning for tactical units • Procurement of medical supplies			

4. Staff Planning/Responsibilities

a. Administration. A staff composed of the G-1/S-1, PAO, SJA, and others as needed, is responsible for advising the commander on NLW employment and ROE. Additionally, administrative personnel are responsible for processing all civilian detainees and claims that arise from NLW employment. As a minimum, the staff must consider the following:

(1) SJA/claims planning.

(2) Chaplain.

(3) Casualty notification.

(4) Detainee disposition plan.

(5) PAO/media planning.

(6) Information dissemination.

b. Intelligence. The G-2/S-2 office is responsible for conducting the IPB process and providing information on the adversary's culture, political stability, and level of dedication to the cause, motivation, leadership, capabilities, equipment, and tactics. The G-2/S-2 should locate native language speakers to assist the interrogator-translator team.

(1) Background on country and culture.

(2) Route/map reconnaissance.

(3) Place sensors for advance notification.

(4) Information on leaders.

(5) LP/OP.

(6) Determine reason for the crowd.

(7) Identify threat.

(8) Analyze traffic.

c. Operations. The G-3/S-3 office is responsible for unit training, nonlethal certification of units, mission analysis, maneuver plans, and CA planning when required.

(1) Tactically integrate NLW and lethal fires.

(2) Fire support plans to include using RCA.

(3) Allocate NLW munitions.

(4) Plan engineer barrier.

d. Logistics. The G-4/S-4 office is responsible for logistics considerations, ship-to-shore movement of weapons and ammunition, storage and maintenance of equipment, resupply, transportation ashore, and other support requirements inherent in using NLW.

(1) Distribute NLW supplies.

(2) Security of rear area facilities.

(3) Resupply NLW munitions.

(4) Special maintenance requirements for NLW equipment and munitions.

e. Communications Officer. The G-6/S-6 staff is responsible for installing, operating, and maintaining communications from units confronting belligerents to the tactical operation center (TOC) or combat operations center (COC). The staff must plan for the effects NLW will have on communications to include proper frequencies.

f. PAO. The PAO provides liaison with outside media and participates in the staff planning process. PAO is also responsible for ensuring the NLW media plan and news releases consider the perceptions of the native media and belligerents or adversaries. PA personnel must receive training on NLW technologies, employment and capabilities, acute and long-term effects, impact on the environment, and legal approval procedures (at a minimum) so they can respond to media and public inquiries.

g. Medical. Medical personnel must be aware of the effects of NLW. Medical personnel should also be familiar with and able to provide decontamination of RCA as well as administer antidotes in cases of adverse reactions to riot control and other agents.

(1) MEDEVAC planning.

(2) Procuring medical supplies.

(3) Hospital planning.

(4) Medical planning for tactical units.

h. Chaplain. The commander should consider chaplain support for counseling and/or religious services for military and civilian personnel. Additionally, chaplains should provide information regarding the religious and cultural background of the local populace.

i. CA. The CA office is responsible for planning and coordinating all aspects of CA activities and providing assistance in gathering intelligence from civilian sources.

(1) Handle displaced civilian personnel.

(2) Coordinate plans for control of civilians.

j. PSYOP. Tactical PSYOP forces serve as an ideal tool for the commander to resolve issues of armed resistance and local interference with military operations while encouraging dialogue and cooperation with noncombatants. This force provides dissemination of PSYOP products (loudspeaker messages, leaflets, posters, etc.) and conducts face-to-face communications with target audience. PSYOP forces provide the supported commanders with nonlethal leverage by influencing foreign populations and expressing information subjectively to influence attitudes and behavior, thereby obtaining compliance or noninterference. They provide public information to foreign populations to support humanitarian activities, ease suffering, and restore or maintain civil order. They serve as the supported commander's voice to foreign populations to convey intent, countering enemy propaganda, misinformation, and opposing information to correctly portray friendly intent and actions. Finally, PSYOP forces deny others the ability to polarize public opinion and political will against the United States and its allies within an AO.

5. Tactics

The commander should include the following considerations in preparing for unit tactical operations involving NLW:

a. Avoid reconfiguring small-unit organization, except for special configurations such as—

(1) Special teams.

(2) Attachments.

(3) Crowd control formations.

(4) Security organizations.

b. Standardize postures in unit tactical SOP. Take the following steps to add a higher volume of specialty impact munitions and equipment within squads, platoons, or units:

(1) Avoid arming troops involved in apprehending designated personnel with long weapons (i.e., shotguns, rifles) because they can be easily grabbed/pulled away by individuals.

(2) Add weapons systems (shotguns) that provide a greater nonlethal capability. When loaded with nonlethal munitions, the longer reloading time of the M203 grenade launcher is protected by coverage from the high rate of fire from shotgun.

(3) Add nonlethal munitions to existing organic weapons systems (M203 grenade launcher).

(4) Avoid arming troops in the front lines of crowd control formations with long weapons.

(5) Focus riot-control capabilities in specialized reaction forces.

(6) Establish base line NLW postures.

(7) Establish command relations for special units.

(8) Maximize distance and barriers.

(9) Create NLW range cards for static positions.

(10) Provide the force with available OC/RCA consistent with ROE.

(11) Always mutually support forces utilizing NLW with lethal means.

(12) Consider environmental conditions (heat/cold/wind) and their effect on the performance of nonlethal munitions. Proper storage, periodic inspections, and rotation of NLW ammunition are critical to maintaining the effectiveness and viability of nonlethal munitions.

c. Unit Weapons, Unit Configurations, and Ammunition Considerations.

(1) Individual. Designate individuals as nonlethal shooters. Nonlethal shooters should also carry lethal munitions for use as required. Lethal rounds must be marked and carried in a manner to prevent confusing nonlethal rounds with lethal rounds.

(2) Squad. Squad leaders should dictate who carries nonlethal grenades and maintain their weapons loaded with lethal ammunition. Ideally, the squad should not change its task organization to accommodate nonlethal equipment, but should designate nonlethal shooters.

(3) Patrols. Commanders should not plan a nonlethal patrol, but should plan a combat/security patrol with a nonlethal capability when mission dictates.

(4) Static Positions. Individuals in static positions should have their weapons loaded with lethal munitions (mission, enemy, terrain and weather, troops and support available–time available and civil considerations dependent) (METT-T/METT-TC). Additionally, they should carry nonlethal munitions. Both shotgun and M203 work well from static positions and should complement each other in a nonlethal mode.

(5) Recovery Teams. Recovery teams have the primary responsibility of retrieving and controlling designated personnel from the crowd; the commander should establish a recovery team before deploying. The force should train team members in unarmed self-defense, open-hand control, and hand cuffing/flex cuffing. It should equip recovery teams with personal protection gear, 9-mm weapon, and flex cuffs. Due to the physical nature of recovery teams, it is important not to arm team members with long rifles. If security for the recovery team is an issue, the commander should attach a security element.

(6) Crowd Control Formations. The unit should establish crowd control formations within a minimum response time. Because of the physical nature of crowd control, individuals in these formations, if possible, should not carry long rifles. These formations should have inherent nonlethal capability. The unit must provide lethal coverage for the entire formation.

d. Individual Elements of a Formation. These elements are: base element, support element, and command element. The formation needs to provide versatility. Restrictions must never be set on numbers, equipment, or changes to the standard formations.

(1) Base Element. Role of the base element (front line):

(a) Base element serves as the front line or the base/foundation of the formation (can be divided into two or three sections or ranks).

(b) Front line consists of the shield holders.

(c) Angle the bottom of the shield to channel flammable liquids away from the body.

(d) Distance or interval between the shields differs based upon the range to the crowd. The closer the crowd, the tighter the interval between the shields.

(e) Provide frontal security.

(f) Hold a solid line.

(g) Provide intimidation factor.

(h) Fire team leaders are in the line holding shields.

(i) Secondary (tertiary) line(s) can consist of the baton holders and/or nonlethal gunners.

(j) Baton holders will—

- Strike with batons.
- Spray OC or disperse smoke.
- Serve as spotters for the NLW gunners.
- Control and detain subjects pulled from the crowd.

(k) Nonlethal gunners will—

- Consist of team leader with M203/M16 (provides nonlethal cover with M203 40-mm munitions and lethal cover with M16).
- Position themselves between shields working with baton holders.
- Serve as immediate control for the team.
- Maintain continuous muzzle awareness of the weapon.
- Provide immediate support to the shields.

(2) Support Element. Role and employment of the support element:

(a) Provide the means to employ lethal and nonlethal capabilities.

(b) Can consist of two ranks: M16/M203 (2nd rank) Mossberg 12-gauge shotguns (3rd rank).

(c) Provide security for search and recovery team.

(d) Consists of munitions teams and/or recovery teams.

(e) Munitions teams provide the means to employ lethal and nonlethal capabilities consisting of M16/M203 and Mossberg 12-gauge shotguns.

(f) Recovery teams provide ability to detain and search rioters downed by nonlethal munitions. Recovery teams will—

- Extract vehicles or personnel from the crowd who are in immediate danger.
- Detain and escort downed rioters to rear of the formation.
- Restrain and search detained individuals.
- Communicate with the base element.
- Spray or employ RCA as needed.
- Fill in the gaps in the front line as needed.
- Insert into a confined or congested area too large for the formation.
- Use a soft base line that capitalizes on the munitions fired to deploy forward of the shields in order to drive the crowd back, retrieve personnel/gear, and apprehend key riot personnel. Employed in pairs (1 x shield man/1 x shotgun).

NOTE: Refer to Marine Corps Security Force (MCSF) formation examples.

(3) Command Element. Role of the command element is to provide command and signal to the formation and consists of the platoon leader/flight commander/ platoon sergeant/flight sergeant, and squad leaders.

(a) Platoon Leader/Flight Commander, Platoon Sergeant/Flight Sergeant—

- Pass verbal commands.
- Maintain order within formations.
- Gauge and evaluate use of munitions and resupply considerations.

(b) Squad Leaders—

- Effectively control the squad.
- Pass information up the chain.
- Employ OC as directed.

(c) Designated Marksman (DM). During a nonlethal engagement, using DM provides confidence and safety to those facing a riot. The DM in an overwatch position scans the crowd, identifies threats and designated personnel for recovery, and fires lethal rounds if a lethal threat is presented. Additionally, DM are ideally suited for flank security and counter sniper operations.

6. Employment Objectives

a. Innovative employment methods of NLW and ingenuity take on critical significance. NLW are employed in concert with other nonlethal and lethal systems to create a synergistic effect.

b. The measures for countering many nonlethal options are usually apparent, quickly learned, and readily available. Because they are not intended to kill, nonlethal options "teach" an adversary what to avoid in the future. Small unit commanders must prepare to keep one step ahead of motivated belligerents.

c. Many nonlethal munitions have both maximum effective and minimum safe ranges. Individuals short of the minimum safe range may suffer severe injuries or death while the effects of most nonlethal munitions are greatly decreased at longer ranges. To be effective, an adversary must be engaged within the optimum range of the munition. Figure V-1 illustrates the ranges for a variety of nonlethal munitions and systems.

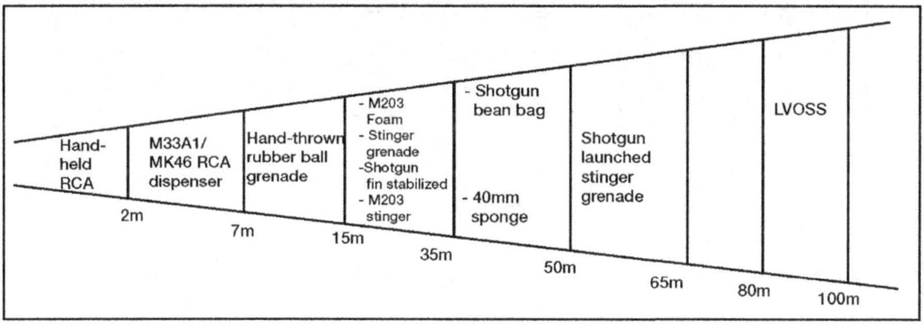

Figure V-1. Nonlethal Weapons Ranges

7. Rules of Engagement

a. U.S. forces follow directives issued by competent military authority that delineate the circumstances and limitations under which deadly force will be initiated against other forces. Intrinsic to the effective use of nonlethal options is the authorization to employ them. ROE are the primary means by which the chain of command can provide guidance to deployed forces in peacetime for handling crises and, in wartime, to help control the level of hostilities.

b. The commander should distinguish ROE from tactical control measures, threat conditions, and arming orders/weapons conditions. Although arming orders are sometimes listed on the same quick reference card as ROE for easy reference, the commander should distinguish ROE from other policies and directives developed, such as weapons confiscation rules.

c. The commander should delegate the decision to use NLW against an adversary to the lowest possible level, preferably to the platoon, flight, or squad level. However, this requires that all personnel, not just leaders, have a clear understanding of ROE and his intent.

d. Using lethal force, employed under the standing ROE, will never be denied. At no time will forces be deployed without the ability to defend themselves against a lethal threat nor will they forego normal training, arming, and equipping for combat. Nonlethal options are a complement to, not a replacement for, lethal force and seek to expand a proactive response across the range of military operations.

e. The following figures are examples of service-specific formations. These examples are not all-inclusive, and the tactical situation will always dictate actual unit employment.

FM 19-15, *Civil Disturbance*; MCWP 3-11.1, *Rifle Company;* and MCWP 3-11.2, *Marine Rifle Squad* provide extensive detail about small-unit formations, their use, and other considerations. Figures V-2 to V-7 are from I and II Marine Expeditionary Force (MEF) Special Operations Training Groups (SOTG) and represent a notional infantry platoon. Figures V-8 to V-13 represent a notional U.S. Army infantry platoon. Figures V-14 to V-21 are additional platoon formations submitted by Marine Corps Security Force Training Company.

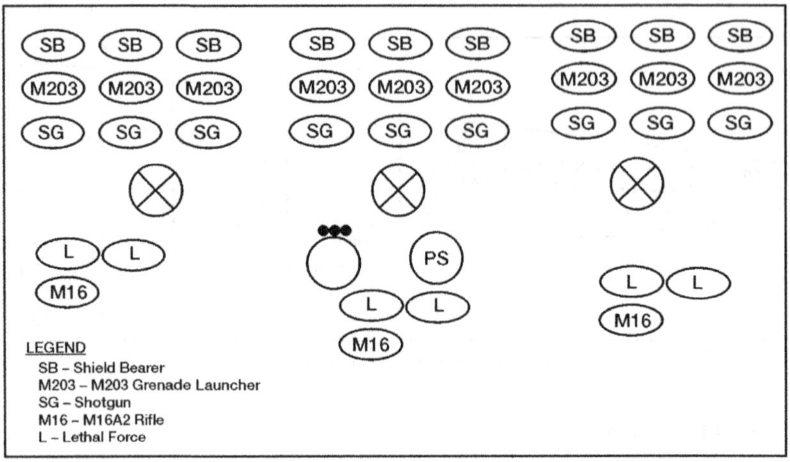

Figure V-2. Platoon in Depth

NOTE: The platoon is the smallest unit capable of independent maneuver. The platoon in depth formation shown above provides the following benefits:

- In-depth support with three ranks.
- Complementary support with the shotguns covering the slower reload time for the M203.
- Additional personnel from each squad to form the recovery teams.
- Each squad is capable of moving as an independent formation forward of the platoon line, if situation dictates.
- Other tactical formations are applicable—echelon, wedge, column.

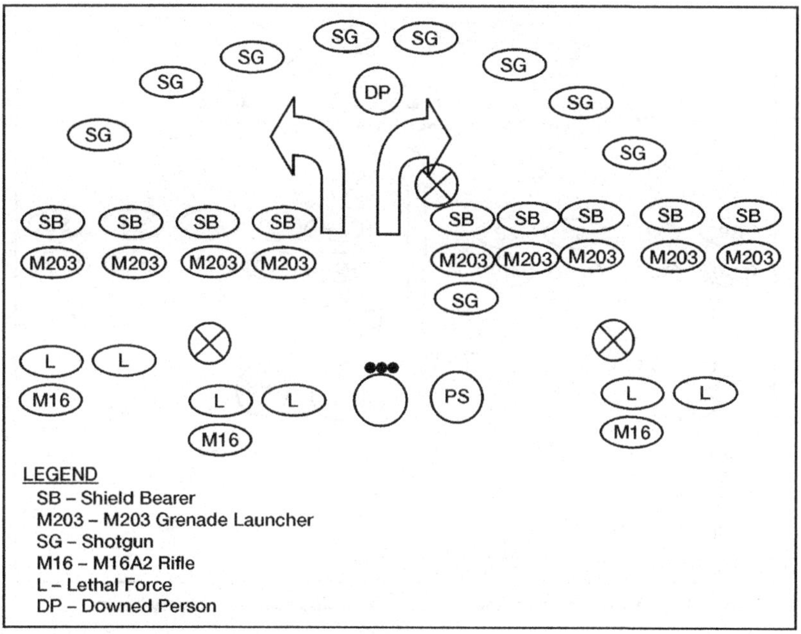

Figure V-3. Platoon Recovery Team Deployment—Security Element

---

NOTE: Security element deploys around downed person.

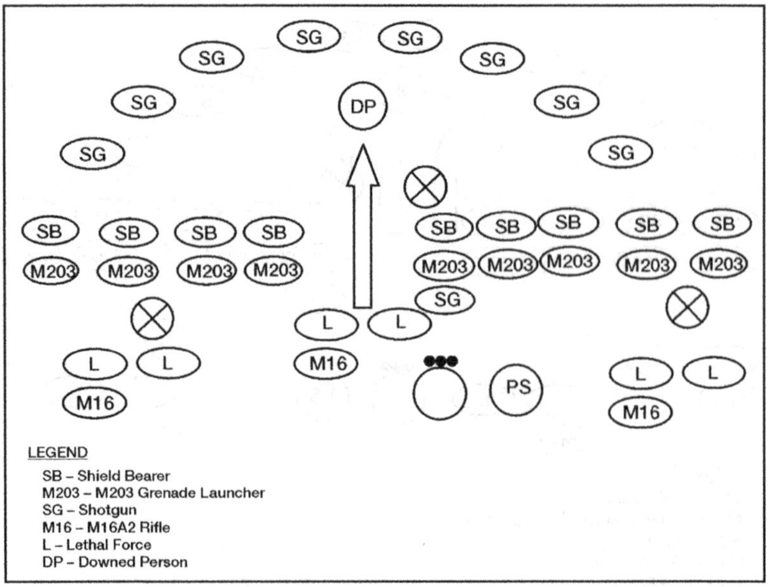

Figure V-4. Platoon Recovery Team Deployment—Recovery Team

NOTE: Recovery team deploys to detain, search, and recover downed person.

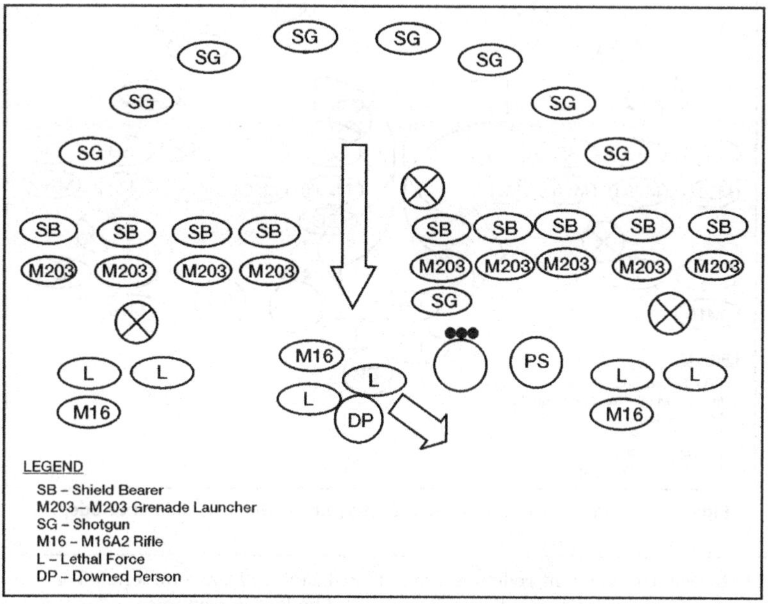

Figure V-5. Platoon Recovery Team Deployment—Recovery Team Redeploy

---

NOTE: Recovery team redeploys with detained person.

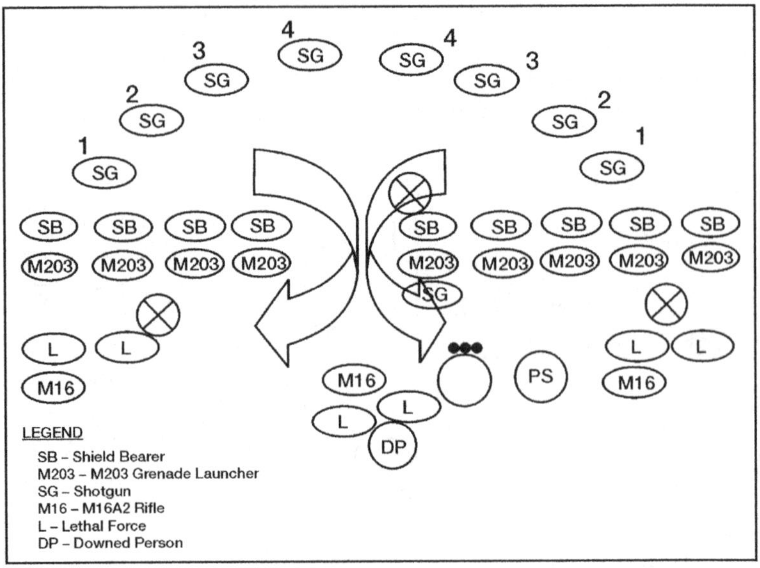

Figure V-6. Platoon Recovery Team Deployment—Security Element Redeploy

NOTE: Security element redeploys back to 3rd rank "by the numbers." Team leader calls "4s in," then "3s in," etc. The last person in calls "last man in" and formation closes the gap.

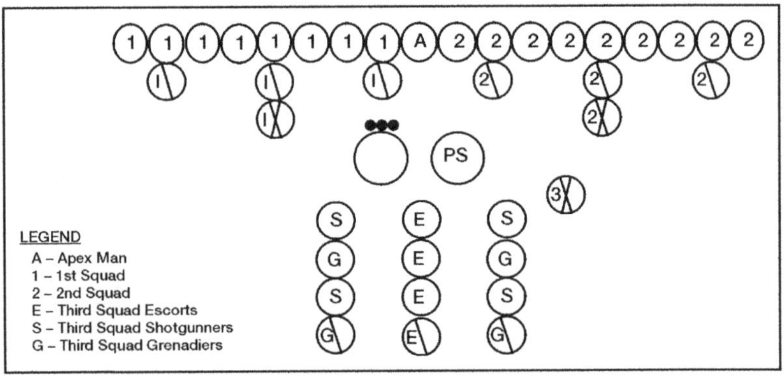

Figure V-7. Platoon On Line—Munitions Squad in Close Support to Cover Flanks

NOTE: Platoon in close support. NLW and lethal weapons are sited to cover the flanks of the formation, as well as support the front line.

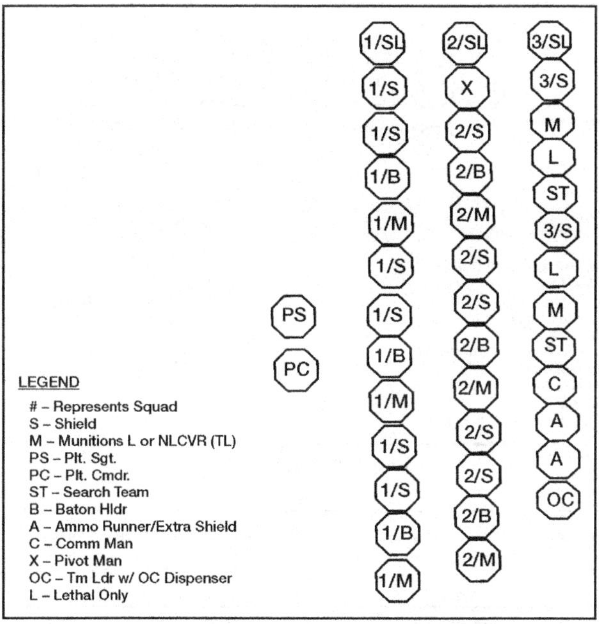

Figure V-8. Platoon Column

NOTE: Platoon in column formation. For administrative movement only.

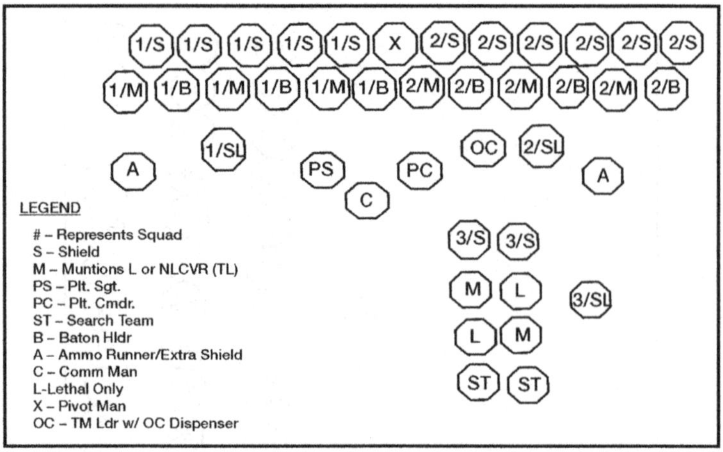

Figure V-9. Platoon On Line

NOTE: Platoon on line—basic formation for the employment of NLW.

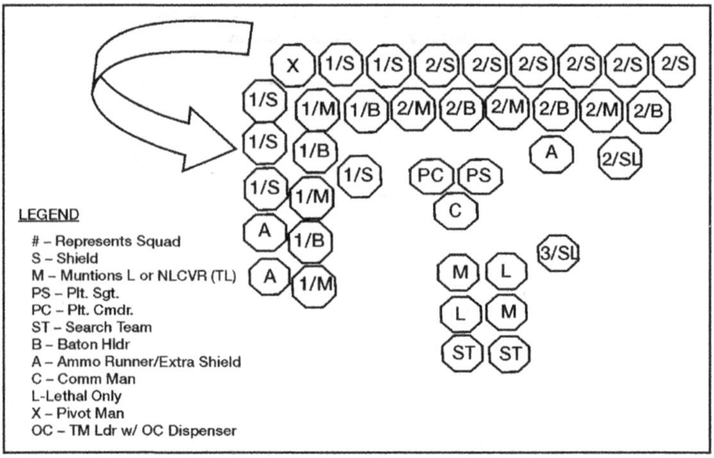

Figure V-10. Platoon Lateral Support, Left (Right)

NOTE: Platoon in lateral support to protect flanks.

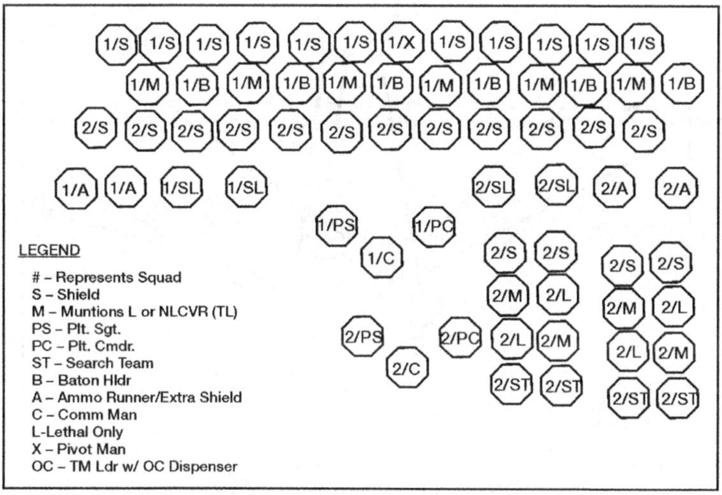

Figure V-11. Two Platoons in Close Support

NOTE: Two platoons in close support—tied in for mutual advantage.

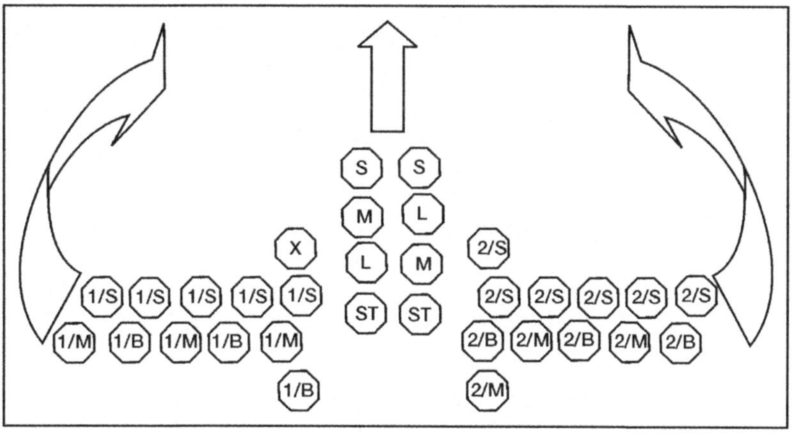

Figure V-12. Recovery Team Deployment

NOTE: Recovery team deploys from formation—remainder of platoon can deploy around the recovery team from either flank.

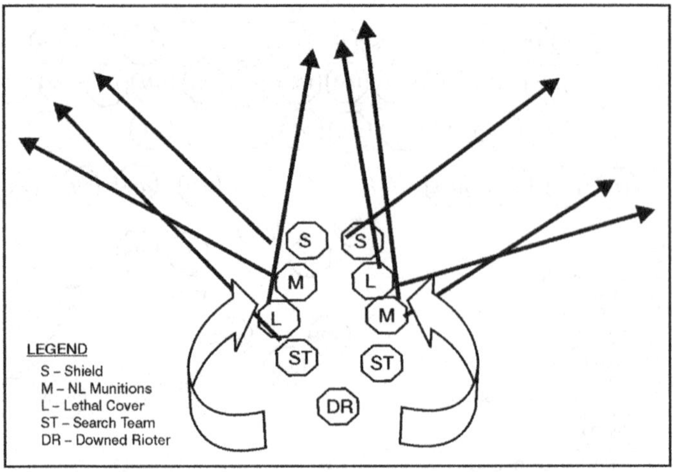

Figure V-13. Recovery Team Actions and Security Posture

***NOTE*:** Recovery team forms security perimeter around downed rioter, conducts search and detain actions. Once actions are complete, recovery team redeploys.

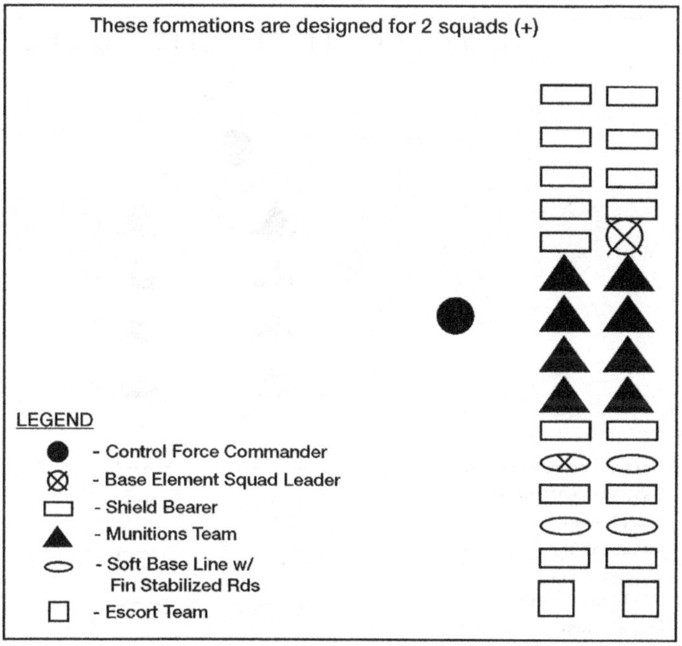

Figure V-14. Platoon in Column (USMC)

NOTE: These formations are designed for 2 squads (+). This allows the platoon to move rapidly when there is no chance of being engaged with the crowd.

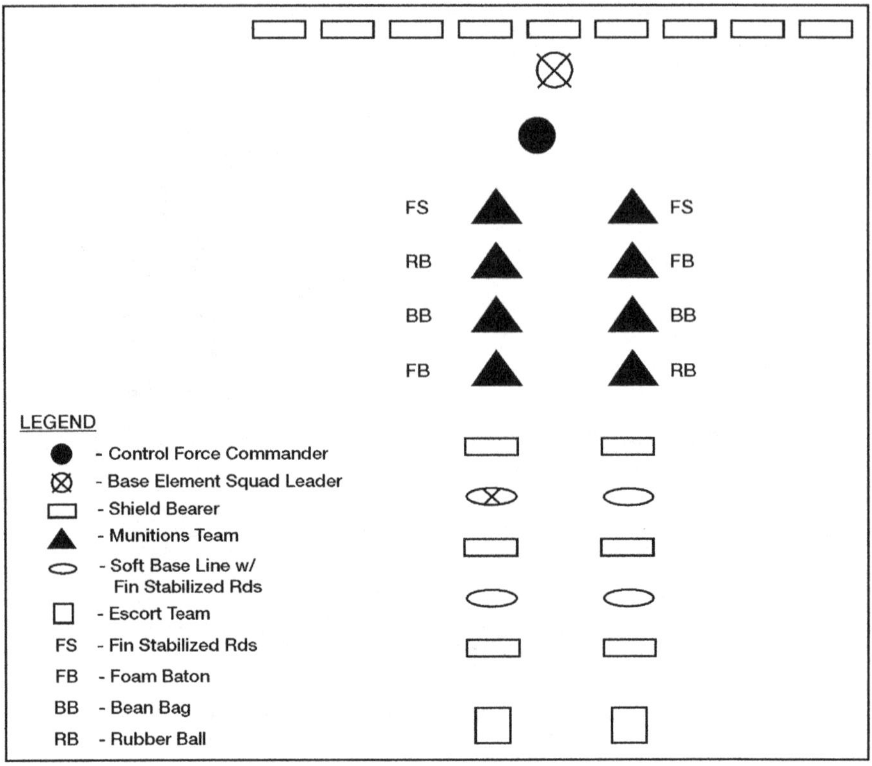

Figure V-15. Platoon in General Support (USMC)

NOTE: The control force commander moves into this formation approximately 60 to 70 meters from the crowd, as the crowd could start to engage the control force with thrown objects.

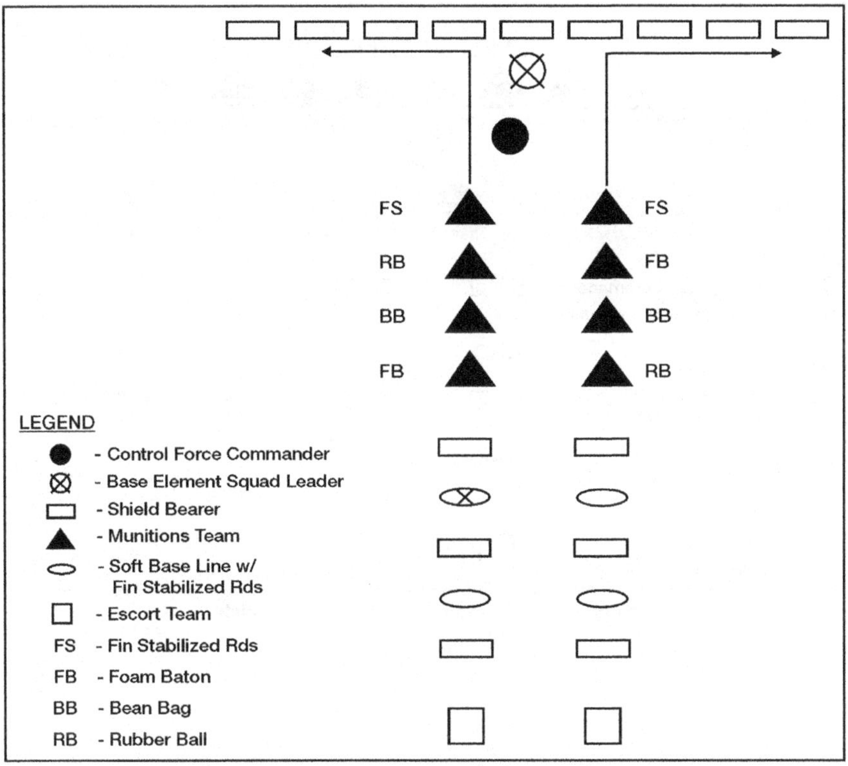

Figure V-16. Platoon in Close Support (Start) (USMC)

NOTE: The control force commander moves into this formation when he feels ready to engage the crowd with nonlethal munitions.

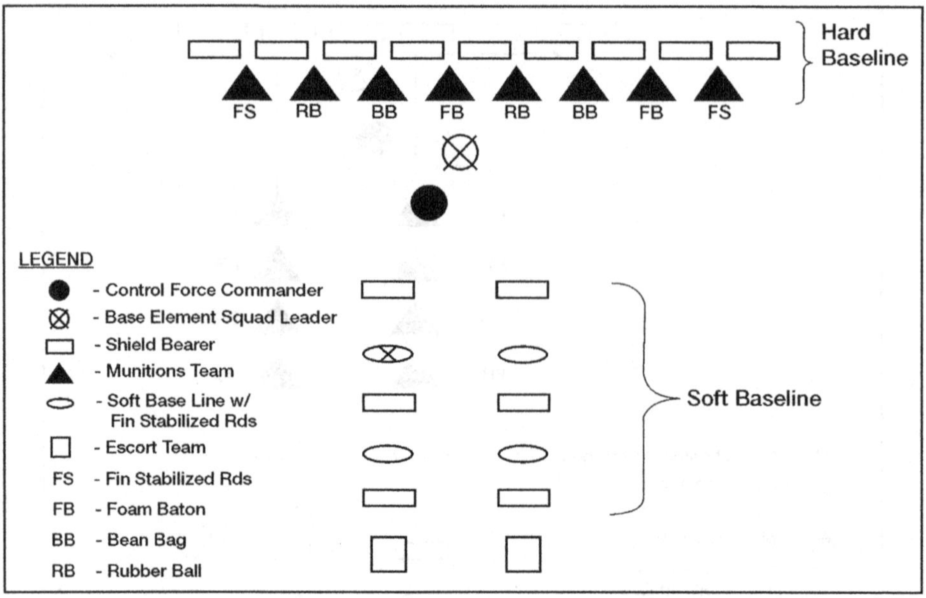

Figure V-17. Platoon in Close Support (Finish)

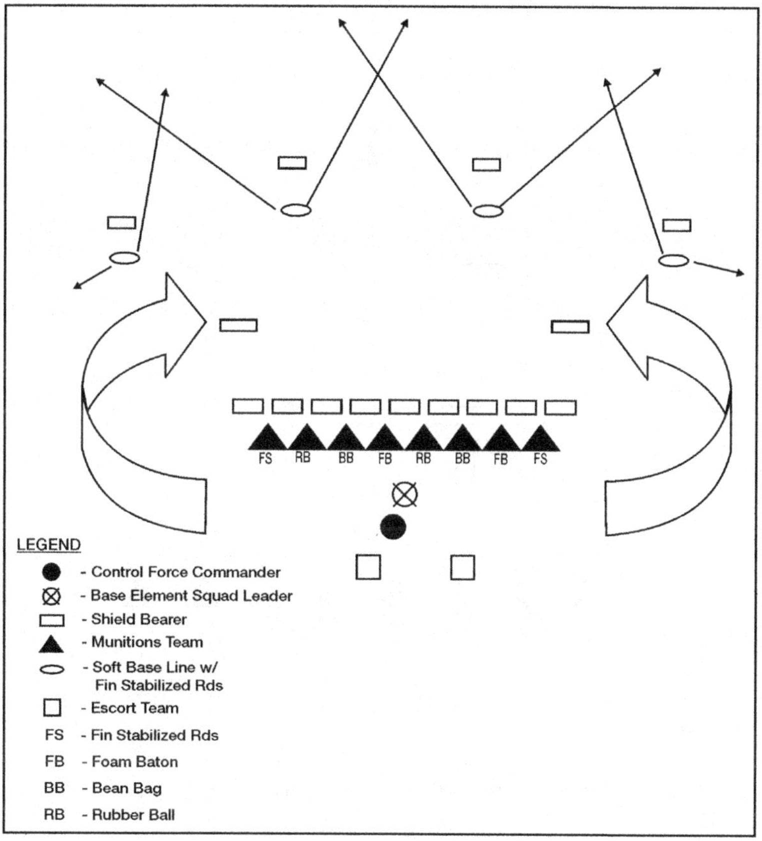

Figure V-18. Deploying Soft Baseline

***NOTE*:** The soft baseline affords the control force commander much flexibility. The primary mission of this formation is to disperse the crowd back using speed and extreme violence of action. By deploying the soft baseline, the commander is able to maneuver the hard baseline where needed (around the barrier plan or obstacles.) The soft baseline is deployed by a predetermined signal (bullhorn siren, whistle blasts, vehicle horn) and will be retrieved using the same signal. This signal conditions the crowd to expect violence of action and causes them to disperse. The soft baseline should never be deployed into a fresh crowd; the crowd needs to be distracted by nonlethal munitions. The formula found to work is distract with stingball grenades, followed by a volley of shotguns and M203 munitions, then send the soft baseline out to disperse or move the crowd back. (Sting them, shoot them, soften them up.)

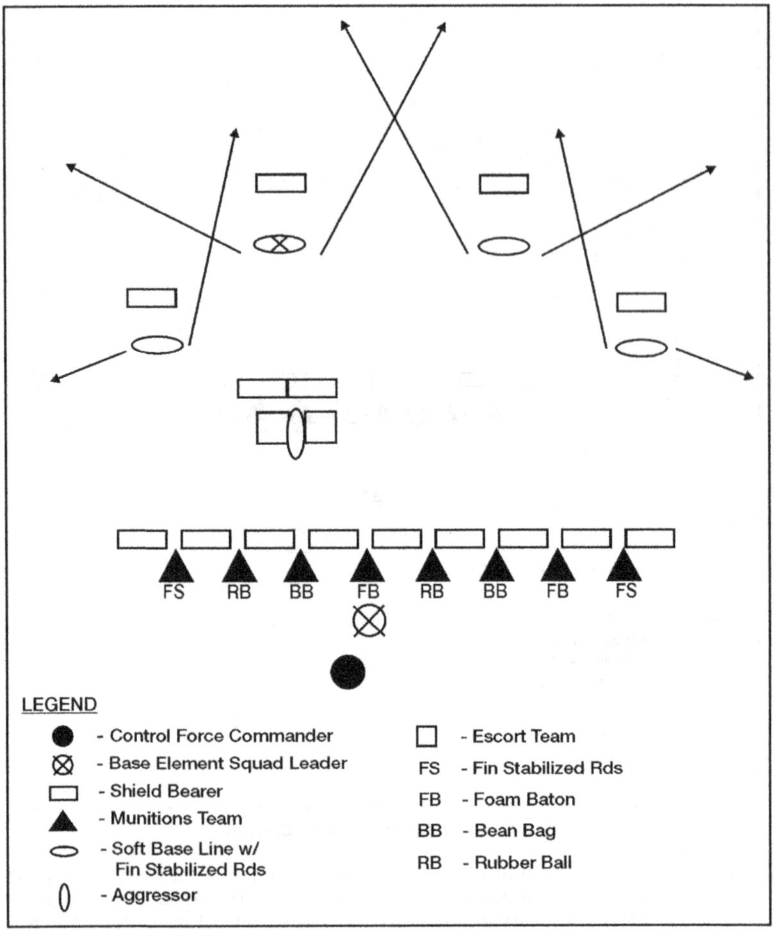

Figure V-19. Deploying Escort Team (Start)

***NOTE*:** When the control force commander identifies the aggressor to retrieve, the command is, "Escort team left/right," then the signal. The signal is the same to deploy the soft baseline. As the soft baseline deploys, the two shields without gunners provide protection for the escort team. The shields remain with the escort team until the soft baseline is called back in, at which time the shields go back to the soft baseline.

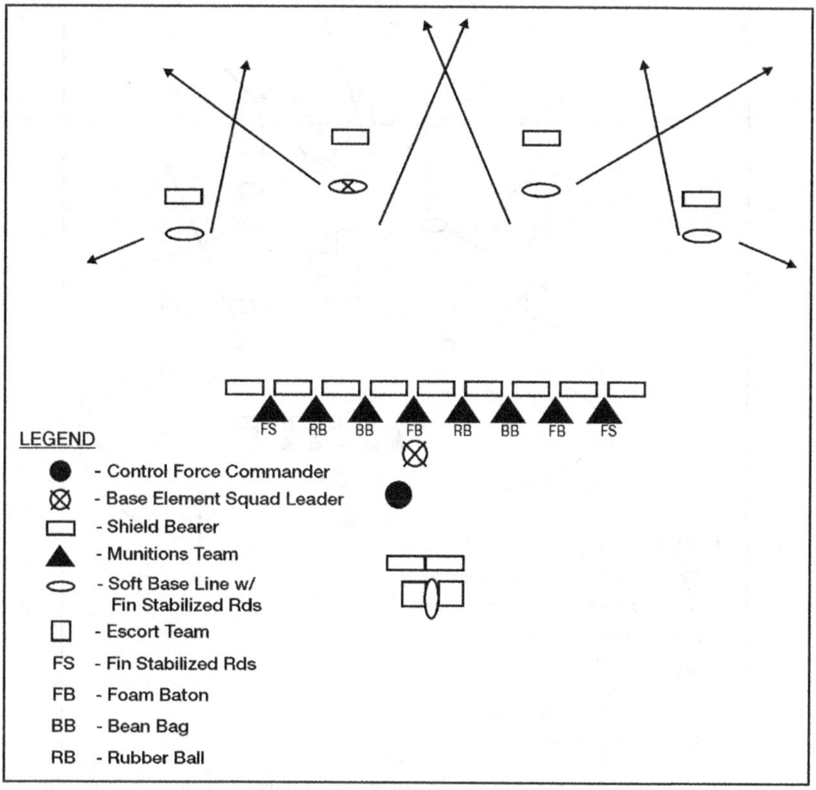

Figure V-20. Deploying Escort Team (Finish)

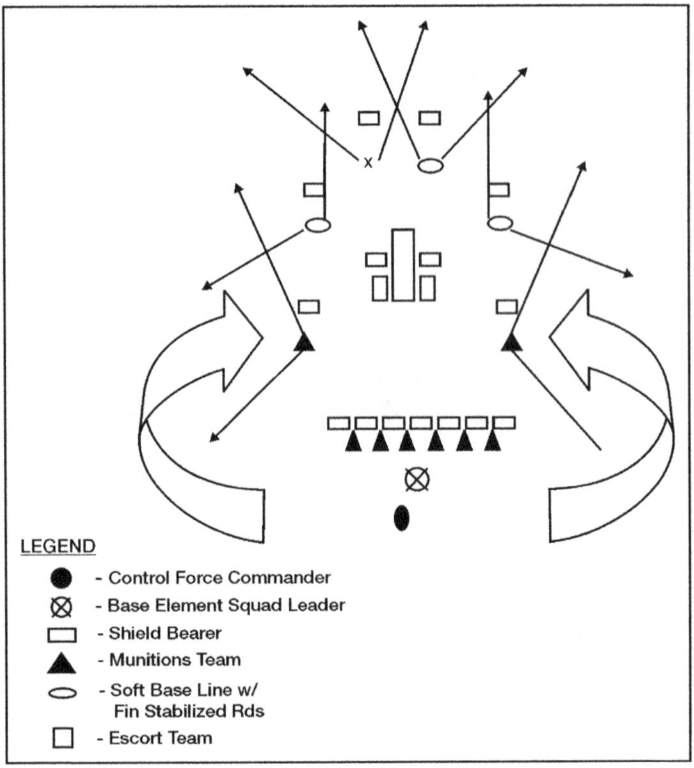

Figure V-21. Recovery of Downed Personnel or Stranded Vehicle (Big Johnson)

NOTE: In the event the crowd attacks personnel or a vehicle, the control force commander can call for a "Big Johnson." In a Big Johnson, the soft base-line deploys as usual except now the shield and shotgunners on the flank of the hard baseline also deploy creating a 190-degree perimeter around the personnel or vehicle. This maneuver must be executed quickly and with extreme violence of action because the crowd already has its hands on individuals in the force. The escort team deploys with the Big Johnson in order to escort or carry injured individuals. The shields without gunners provide protection for the escort team.

Chapter VI

LESSONS LEARNED

1. Background

Although lessons learned from employing NLW are still being studied and impacts analyzed, it is beneficial to look at the preliminary lessons gained from the NLW experience. These lessons are far from conclusive, because U.S. forces have yet to see the first large-scale employment of NLW. However, U.S. forces can gain tremendous insight from the experience of others and use what has been learned to guide future NLW employment.

2. Lessons Learned

a. Force Continuum. Using nonlethal options needs to be based on the principle of a graduated and proportional response, using the decisive force necessary to respond to the threat.

b. Attempts to legislate and regulate every aspect of NLW employment can prove disadvantageous to the on-scene commander. During the evacuation of U.N. forces in Somalia (OPERATION UNITED SHIELD, 1995), the ROE made an arbitrary and unnecessary distinction between situations justifying deadly force and those that justified only lesser means. Crowd control items were restricted to situations requiring deadly force. Many soldiers, airmen, and Marines asked, "why resort to nonlethal means at all if I have to wait until deadly force is authorized before I can shoot them with a beanbag?" Additionally, the use of sticky foam was limited to deadly force situations because of the perception that adversaries might suffocate from its application. This stipulation effectively eliminated sticky foam as a nonlethal option to the forces in Somalia.

c. With ROE that take into account the force continuum and encourage graduated response to nonlethal threats, the commander can employ NLW to enhance his flexibility, deterrence ability, and reaction to uncertain situations. To accomplish the mission with minimum loss of life and property, commanders need to have ROE that do not restrict them from properly employing NLW.

Lessons Learned from OPERATION JOINT GUARDIAN 1999 (Battalion Landing Team (BLT) 3/8 26 Marine Expeditionary Unit (MEU) Peace Operations in Kosovo:

ROE significantly impaired the employment of NLW. For example, OC was not authorized because it was considered a chemical agent. Nonlethal direct fire munitions were not authorized during the initial occupation of the province. This significantly hampered the individual Marine's ability to neutralize a threat where deadly force would not have been justified but action was required. Overly restrictive ROE thus negated the flexibility of response that NLW are supposed to provide. The intent of deploying NLW is to provide the operator the capability and flexibility to neutralize a threat without having to apply deadly force. Thus, the ROE for NLW should ideally be less restrictive, not the same or more restrictive, than those for lethal force.

It is often enticing for authors of ROE involving the use of NLW to base those ROE on the principle of minimum force necessary. This may be appropriate for base and station or homeland security ROE; however, for expeditionary operations the decision to mandate minimum force deserves critical consideration. Commanders in military operations have traditionally gained advantage over their enemy not only by the actual combat power that they bring to the fight, but also from that enemy's perceptions and beliefs about the intensity and resolve with which that combat power will be applied. When ROE are written stating that the confrontations will be addressed using the minimum force necessary, this advantage is lost. ROE developed for expeditionary nonlethal operations should be based on the concept of decisive and compelling force, rather than the minimum force. This will maintain an adversary's hesitancy to harass and endanger U.S. forces even after that adversary has tested or probed the response of those U.S. forces. Further, this will allow synergy to be gained through the complimentary employment of lethal and NLW by providing more response options to the small unit leader.

3. Employment

a. Through the use of modeling, simulation, and field exercises, new concepts for employing NLW are continually being developed. The force should train with NLW to the same degree as for lethal weapons. In this way, units become confident when using nonlethal options in chaotic and uncertain situations. The commander should push the decision to employ NLW to the lowest possible tactical level. Commanders who intend to use NLW in achieving military objectives must recognize the fluidity of the situations in which they are likely to be employed. The decisionmaking cycle is short, stressful, and the level of violence can move in both directions along the force continuum. In quick succession, a small unit leader might employ nonlethal munitions, decide to employ deadly force to an immediate specific threat that arises, then return to using nonlethal means in response to unarmed hostile elements.

b. A mixture of nonlethal "tools" has been determined to be advantageous in an operational scenario. As the situation develops, commanders can apply different nonlethal responses with respect to range/distance capability and severity of effect. For example, to disperse a crowd, area fire weapons would precede use of close-in point target nonlethal munitions.

c. An important point to remember is that NLW are just one element of a graduated response available to the force commander. Other factors make up a large contribution to the success of U.S. forces, including barrier plans and PSYOP. An example of this operation would be using barriers such as roadblocks or concertina wire, warning commands to inform a mob of potential responses based on their actions, and only then taking actions through escalated employment of NLW. These continued responses would involve smoke, RCA, and blunt munitions. A specific verbal warning should precede every escalation of nonlethal means.

d. Whenever NLW are employed, a separate lethal capability must always be in place. Troops must always be prepared to protect themselves as required by lethal means, because it is impossible to accurately predict the response of an individual or mob to NLW employment.

Lessons Learned from OPERATION ESSENTIAL HARVEST 2001 (U.S. Embassy Reinforcement operations conducted at the U.S. Embassy Skopje Macedonia by 5th Platoon, 2nd Fleet Antiterrorism Security Team [FAST] USMC):

The anticipated threats were violent demonstrations directed against U.S. facilities in the AO. In this case, USMC forces tasked to reinforce the embassy were equipped with a compliment of NLW. ROE took into account the capabilities and limitations of the NLW. The ROE and the NLW systems on the ground were coordinated and thus compatible. In addition, it was found that integrated employment of different nonlethal munitions greatly enhanced that force's ability to suppress the threats encountered (large rioting mobs). Random security patrols reinforced with military working dogs proved very effective. Reinforcement with multiple M203s provided USMC units with both a lethal and nonlethal capability in the same weapons system. The M203 was particularly useful as it gave the individual Marine the means to quickly transition between a lethal (5.56-mm) and nonlethal (40-mm nonlethal) capability. A basic understanding of the capabilities and limitations of nonlethal munitions enabled a relatively small force (40-50 Marines) to counter a sizable threat many times its size (500-1000 rioters). Although NLW were not used, the capability and preparation of the units provided the adequate deterrence that prevented the situation form escalating.

4. Publicity and Media

a. The commander can use the media to his advantage, as they will inform the local population of the military's capabilities and responses available. During OPERATION UNITED SHIELD in Somalia, the local media was informed of the Marines' nonlethal options and "new technologies" in order to discourage hostile actions. The media was encouraged to report this to the population. Intelligence had indicated the Somalis were planning to mix civilians with armed clansmen, creating hostile mobs. In fact, feuding warlords had previously successfully demonstrated this strategy on several occasions. Lack of deadly mob incidents implies Somalis were not anxious to test this new technology. Evidently, publicity had directly contributed to mission accomplishment.

b. The public dissemination of the presence of NLW capabilities must strike a balance between information sufficient to discourage hostile actions and information that enables the local population to learn too much of the capabilities of NLW. Too much information allows adversaries to develop countermeasures or avoid the NLW effects. Commanders who employ NLW systems must involve the PAO from the beginning of planning. The PAO must be familiar with the objectives of the mission, commander's intent, ROE, NLW and lethal weapons employed, and be prepared to advise the commander on the best methods of dealing with outside media. The commander should fully integrate the PA plan with PSYOP objectives and the concept of operations. Additionally, he must work the media to ensure the "sensationalism" of such a new technology does not create unrealistic expectations where he can be unfairly judged or constrained.

5. Capabilities and Impact on Mission Accomplishment

a. The commander should understand the unique capabilities of NLW and use them to accomplish the mission. He should use NLW in conjunction with the planned employment of the force and within established ROE to minimize loss of life and property while accomplishing the mission. The force requires early hands on training to recognize the capabilities and limitations of NLW. Appendix D has sample training plans developed to meet these objectives.

b. The force commander must fully integrate planning for using nonlethal capabilities as quickly as possible in the staff planning cycle. In order to have a full range of alternatives for use against unarmed but hostile elements, the force needs to deliver nonlethal systems, complete necessary ROE and legal review, and make sufficient quantities of each NLW readily accessible to the force. The choice of nonlethal capabilities should be based on several factors and the choice can be made after addressing these questions:

(1) First, are the NLW systems available and can they be delivered on time?

(2) Second, are they compatible with, and do they complement existing weapons and training?

(3) Third, are there clear and unambiguous guidelines and ROE for employment?

(4) Finally, can the NLW be effectively employed to save lives and contribute to mission accomplishment?

c. Leaders need to anticipate local civilians developing effective countermeasures to their NLW and tactics. Therefore, prior to beginning their mission, leaders must determine and rehearse actions to take when nonlethal munitions are effectively countered.

d. Using nonlethal and lethal force are not mutually exclusive. In many instances, the commander can gain synergistic effects by using a combination of both. The following are some examples of this:

(1) Utilizing NLW to separate agitators from noncombatants with simultaneous employment of lethal force against hostiles or combatants where deadly force is justified.

(2) Utilizing NLW when making entry into structures or rooms to gain time in order to discern temper and intent of occupants prior to applying deadly force.

6. Knowing Your Adversary

a. Although commanders desire thorough intelligence of any potential adversary, such a goal may be difficult to reach in a NLW employment environment. During OPERATION UNITED SHIELD in Somalia and OPERATION UPHOLD DEMOCRACY (1995) in Haiti, a study of the local population's culture, history, and customs enabled commanders to anticipate and control many situations before they got out of hand. On the other hand, during OPERATION UPHOLD DEMOCRACY, poor intelligence and a lack of understanding between the two cultures led to the Haitians and Cubans being lodged together in Guantanamo Bay. This situation led to riots between factions who had a deep dislike for each other. By understanding a particular crowd's motivation, friendly forces may be better able to take necessary steps to prevent escalation resulting in the use of deadly force. Early training in crowd dynamics and riot control will enhance the effective use of NLW.

Lessons Learned from U.S. Peace Operations, 739th MP Battalion (Bn), January 2001

An Army MP platoon received a mission to clear a rioting crowd of 300 civilian locals from a main supply route. The platoon placed vehicle-mounted machine guns in overwatch positions to allow quick transition from NLW to conventional rounds. 12-gauge shotgun teams armed with fin stabilized rounds carried M9 pistols for lethal backup. Two volleys of shotgun and M203 nonlethal rounds successfully dispersed the crowd. The platoon had utilized INIWIC graduates to conduct preparatory training on using NLW. The NLW were extremely effective in pushing back the crowd and causing it to disperse. The 40-mm rubber ball round was moderately effective at close range, but very inaccurate at longer ranges. Shotguns were more effective than the M203 due to their high rate of fire and quick-loading capabilities.

b. Specific lessons learned during this operation are as follows:

(1) Placing interpreters is critical. The solution in this incident was to place interpreters approximately 25 meters behind formation so they could deploy forward as needed. Interpreters were often needed to communicate with crowds and with adjacent forces and coalition units that were organic to Kosovo forces (KFOR) that also responded to the incident.

(2) Leaders as well as soldiers should prepare to speak to the media. The command should have a PAO representative on hand to deal with the media.

(3) The commander should thoroughly brief the transition to lethal force from NLW.

(4) U.S. commanders in Kosovo engaged in the decisionmaking process had disparate levels of knowledge of nonlethal capabilities, constraints, and operational considerations. Therefore, leader training and orientation requires more emphasis.

(5) ROE were vague regarding the use of NLW. In many operations, NLW use is delegated to the on-site commander. Delegating to the lowest possible level is the preferred method.

(6) A unit needs a mixture of nonlethal rounds based on the size of crowd, composition of the crowd, available weapons systems, and synergistic effect of selected nonlethal combinations.

(7) When a RRF/QRF arrives on station with NLW to support a force equipped only with lethal means, the commander should address the following command and control concerns:

(a) Who makes the decision to employ NLW?

(b) Is the reinforced commander familiar with NLW?

(c) Who controls the reinforced unit?

(d) How does this relationship change as the situation develops?

7. Summary

a. The wide range of military operations in which U.S. forces are involved has demanded the introduction of NLW into the arsenal of weaponry available to commanders. NLW provide a graduated and proportional response to threats posed by a variety of adversaries. The ROE must be consistent with the rapidly changing situation of today's chaotic and uncertain military operations. This reality demands a high level of training in properly employing NLW.

b. Effective use of NLW involves a synergy between PSYOP activities, intelligence gathering, and professional media/PA plan. NLW effectiveness is enhanced through integration of these activities to reduce the chance adversaries will be able to develop effective countermeasures. NLW, when combined with lethal weapons, provide the commander with a force response across a wide range of military operations.

Appendix A

DOD Nonlethal Weapons and Equipment Item Description

The following is a cumulative list of NLW and equipment currently fielded within the DOD. Each service has various components of the following list in its inventory. A list of the service's items is located in Appendix E.

 **Figure A-1. Riot Face Shield**	1. Riot Face Shield: The riot face shield is a non-ballistic face shield designed to protect the service member's face from objects thrown or attacks from non-ballistic weapons. (Figure A-1)
No picture available.	2. Ballistic Face Shield: The ballistic face shield provides the service member with face protection from a wide range of threats (i.e. debris, liquids, handthrown objects). The shield is manufactured from acrylic and bullet resistant materials to provide Level IIIA protection (9-mm and .44 magnum pistol rounds), weighs 3.4 pounds and is compatible with the PASGT helmet and M-17/M-40 protective masks.
 **Figure A-2. Riot Body Shield**	3. Riot Body Shield: The riot body shield is a non-ballistic body shield designed to protect the service member's body from objects thrown or attacks from non-ballistic weapons. (Figure A-2)
No picture available.	4. Ballistic Body Shield with Light Kit: The ballistic shield provides the service member with torso protection from a wide range of threats (i.e. debris, liquids, hand thrown objects). The shield is manufactured from Spectra Shield resistant material painted anti-reflective flat black, and is capable of providing Level IIIA protection (9-mm and .44 magnum pistol rounds). The shield weighs 18 pounds and is 20 x 36 inches in size with a view port of 4 x 16 inches.

 **Figure A-3. Riot Shin Guard**	5. Riot Shin Guard: The riot shin guard is a non-ballistic shin guard designed to protect the service member's shins from objects thrown or attacks from non-ballistic weapons. (Figure A-3)
No picture available.	6. Ballistic Shin Guards: Ballistic shin guards provide the service member with leg protection from a wide range of threats (i.e. debris, liquids, hand thrown objects). The guards are manufactured from Kevlar KM2 and provide Level IIIA protection (9-mm and .44 magnum pistol rounds). They are available in three sizes and weigh between 7 to 10 pounds.
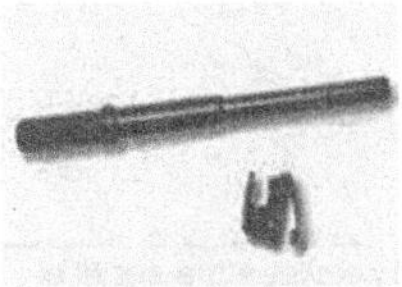 **Figure A-4. Expandable Riot Baton (with holster)**	7. Expandable Riot Baton (with holster): The expandable riot baton provides the service member a nonlethal means of crowd control and self-defense. The baton comes with a mounting device that attaches the baton to a belt. The baton is 24 to 36 inches in length. (Figure A-4)
No picture available.	8. Wooden Baton: The wooden baton provides self-defense for the service member and the option between physical force and lethal force. The baton comes with a mounting device that attaches the baton to a belt. The baton is 36 inches in length.
No picture available.	9. Practice Riot Batons: The practice riot batons provide the service member with a safe dynamic training environment for properly using a riot baton. The batons are available in a variety of lengths.
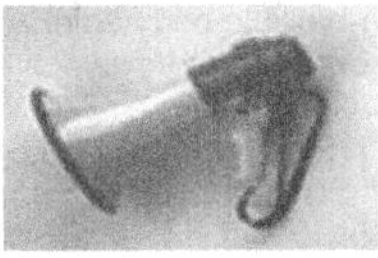 **Figure A-5. Portable Bullhorn**	10. Portable Bullhorn: The portable bullhorn provides the service member the capability to control his forces by projecting his voice above the noise and commotion created by crowds and mobs. (Figure A-5)
No picture available	11. Ground-Mounted Public Address System: The ground-mounted public address system is a communication enhancement device for conducting crowd control tactics. The system facilitates communication with the crowd in the absence of PSYOP support and assists with communication of commands to troops engaged in crowd-over-crowd noise.
No picture available.	12. M7 Individual Voice Amplification System: The M7 individual voice amplification system is a communication enhancement device for conducting crowd control tactics while wearing the M40 protective mask and using riot control agents. The M7 is fitted to the M40 protective mask to facilitate oral communication and increase the user's ability to communicate using radios and other devices.

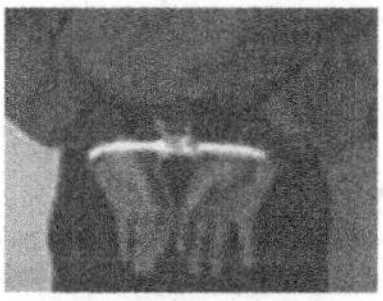

Figure A-6. Disposable Handcuff

13. Disposable Handcuff: The disposable handcuff provides service members with a means of restraining and controlling suspects, prisoners, and other individuals encountered during crowd control. The disposable handcuff is designed for one-time use. (Figure A-6)

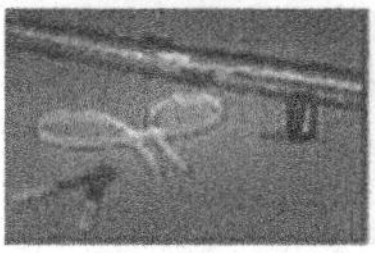

Figure A-7. Training Disposable Handcuff

14. Training Disposable Handcuff: The training disposable handcuff provides service members with a means for training in handcuffing procedures prior to actual use. The training disposable handcuff is red and reusable. (Figure A-7)

Figure A-8. MK-4 Pepper Spray Pouch

15. MK-4 Pepper Spray Pouch: The MK-4 pepper spray pouch is designed to carry the MK-4 pepper spray canister. (Figure A-8)

Figure A-9. MK-4 Individual Live OC Pepper Spray

16. MK-4 Individual Live OC Pepper Spray: The MK-4 individual pepper spray has a range of up to 12 feet in a target-specific stream. It contains 30 to 35 half-second bursts. (Figure A-9)

No picture available.

17. M-36 Individual Riot Control Disperser: M-36 Individual Riot Control Disperser has a range of up to 12 feet in a target-specific stream. It contains 30 to 35 half-second bursts. It is loaded with either CN or CS agent.

Figure A-10. MK-9 Pepper Spray Pouch

18. MK-9 Pepper Spray Pouch: The MK-9 pepper spray pouch is designed to carry the MK-9 pepper spray canister. (Figure A-10)

Figure A-11. MK-9 Squad Live OC Pepper Spray

19. MK-9 Squad Live OC Pepper Spray: The MK-9 squad pepper spray has a range of up to 15 feet in a target-specific fog. It contains 8 to10 half-second high volume bursts. (Figure A-11)

Figure A-12. MK-46 Platoon Live (Horizontal) OC Pepper Spray

20. MK-46 Platoon Live (Horizontal) OC Pepper Spray: The MK-46 platoon pepper spray has a range of up to 25 to 30 feet in a target-specific dispersed spray pattern. It contains up to 12 one-second high volume bursts. (Figure A-12)

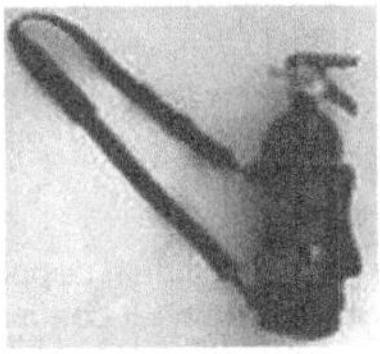

Figure A-13. MK-46 Sling

21. MK-46 Sling: The MK-46 sling gives a service member the ability to sling the MK-46 canister on the shoulder. (Figure A-13)

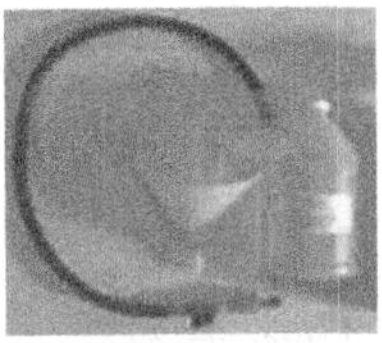

Figure A-14. MK-46 Hardware Kit

22. MK-46 Hardware Kit: The MK-46 hardware kit consists of a transfer tube, funnel, and transfer bottle. A compressed nitrogen source is also required to utilize the kit. (Figure A-14)

Figure A-15. MK-46 Refill Live OC Pepper Spray

23. MK-46 Refill Live OC Pepper Spray: The MK-46 refill solution comes in a one-gallon container. (Figure A-15)

Figure A-16. MK-4 Individual Inert OC Pepper Spray

24. MK-4 Individual Inert OC Pepper Spray: The MK-4 inert individual pepper spray has a range of up to 12 feet in a target-specific stream. It contains 30 to 35 half-second bursts. It provides the service member with an inert canister for familiarization and training purposes. The inert device is loaded with a non-irritant, non-toxic formulation payload. The device may also contain a non-toxic, water-soluble marker to allow immediate feedback on content delivery to target. The MK-4 device is marked "Inert." (Figure A-16)

Figure A-17. MK-9 Squad Inert OC Pepper Spray

25. MK-9 Squad Inert OC Pepper Spray: The MK-9 squad inert pepper spray has a range of up to 15 feet in a target-specific fog. It contains 8 to 10 half-second high volume bursts. It provides the service member with an inert canister for familiarization and training purposes. The inert device is loaded with a non-irritant, non-toxic formulation payload. The device may also contain a non-toxic, water-soluble marker to allow immediate feedback on content delivery to target. The inert MK-9 device is prominently marked "Inert" with a yellow sticker attached. (Figure A-17)

Figure A-18. MK-46 Platoon Inert OC Pepper Spray

26. MK-46 Platoon Inert OC Pepper Spray: The MK-46 platoon inert pepper spray has a range of up to 25 to 30 feet in a target-specific dispersed spray pattern. It contains up to 12 one-second high volume bursts. It provides the service member with an inert canister for familiarization and training purposes. The inert device is loaded win a non-irritant, non-toxic formulation payload. The device may also contain a non-toxic, water-soluble marker to allow immediate feedback on content delivery to target. The inert MK-46 device is prominently marked "Inert" with a yellow sticker attached. (Figure A-18)

Figure A-19. MK-46 Refill Inert OC Pepper Spray

27. MK-46 Refill Inert OC Pepper Spray: The MK-46 refill solution comes in a one-gallon container. (Figure A-19)

Figure A-20. FIST Training Suit

28. FIST Training Suit: The FIST training suit is made of closed-cell shock-absorbent foam that allows service members to hone their riot baton skills through realistic hand-to-hand engagement training. The FIST suit can absorb blows inflicted by the current expandable and wood riot baton. It provides protection for the head, face, hands, and legs, without significant degradation of the wearer's mobility. The FIST suit consists of a helmet, chest and back protector, bicep and forearm protectors, groin and buttock protectors, thigh and shin protectors, gloves, and carry bag. (Figure A-20)

NOTE: FIST is the current manufacturer of this training suit).

Figure A-21. Riot Training Strike Bag

29. Riot Training Strike Bag: The riot training strike bag gives the service member a strike bag for training and proficiency in open-hand control techniques. (Figure A-21)

Figure A-22. Caltrop

30. Caltrop: The caltrop is a field expedient area denial system. Securing approximately 50 caltrops to 550 military cord best employs it. This allows a service member to quickly toss/recover the system, using the caltrops to deny vehicle/pedestrian movement in a designated area. The four prongs are approximately two inches tall and will puncture tires or boots. (Figure A-22)

Figure A-23. Road Side Spike Strip

31. Road Side Spike Strip: The spike strip is designed to puncture a pneumatic tire as the vehicle makes contact with the spike strip. The spikes pivot into the tire and the tip-guards are pushed down. The spikes are imbedded deeply into the tire. The spikes are pulled out of the unit and remain in the tire. On the second revolution they will be pushed directly into the tire, allowing air to escape slowly without causing a blowout or sudden loss of control. The average tire will pick up 4 to 6 spikes with the deflation time of about 12 to 20 seconds. It allows a service member to quickly toss/recover the system, impeding vehicle movement in a designated area. It allows a force to provide additional physical security in areas where the caltrop cannot be tactically employed. (Figure A-23)

Figure A-24. High-Intensity Search Light

32. The high-intensity searchlight is a rugged, lightweight, hand-held, high-intensity spotlight for use during low-light conditions. Two different high-intensity lights were fielded. The Maxabeam light (shown on top) was fielded in 1997. It was replaced by the Nighthunter light (shown underneath) in 2000. (Figure A-24)

Figure A-25. Individual Light Holster (6P)

33. Individual Light Holster (6P): The 6P individual light holster is designed to hold the 6P individual light. (Figure A-25)

Figure A-26. 6P Individual Light

34. 6P Individual Light: The 6P individual light is issued to service members within the platoon. (Figure A-26)

Figure A-27. 6P Individual Light Accessory Kit

35. 6P Individual Light Accessory Kit: The 6P individual light accessory kit includes a spare light bulb installation kit. (Figure A-27)

Figure A-28. 3V Lithium Batteries

36. 3V Lithium Batteries: Two 3V Lithium Batteries are required to operate the 6P individual light. (Figure A-28)

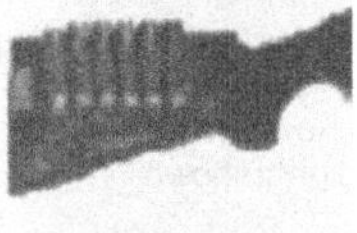

Figure A-29. 12-Gauge Buttstock Cuff

37. 12-Gauge Buttstock Cuff: The buttstock cuff is designed to fasten to the buttstock of most shotguns to hold seven 12-gauge cartridges for quick loading of the shotgun. (Figure A-29)

Figure A-30. 12-Gauge Utility Pouch

38. 12-Gauge Utility Pouch: The 12-gauge utility pouch is designed to hold twenty-five 12-gauge munitions and attaches to the service member's combat gear. (Figure A-30)

Figure A-31. 40-mm Carrying Pouch

39. 40-mm Carrying Pouch: The 40-mm carrying pouch is partitioned in a manner to provide the service member with a means of carrying not less than six 40-mm nonlethal rounds in separate compartments. It attaches to the service member's combat gear. (Figure A-31)

Figure A-32. Diversionary/Rubber Ball Grenade Pouch

40. Diversionary/Rubber Ball Grenade Pouch: The diversionary/rubber ball grenade pouch is an ammunition carrying pouch partitioned in a manner to provide the service member with a means of holding up to six rubber ball grenades or six flash bang grenades in separate compartments. (Figure A-32)

Figure A-33. 12-Gauge Mossberg 500A2/590 Shotgun

41. 12-Gauge Mossberg 500A2/590 Shotgun: The 12-gauge Mossberg 500A2 or 590 shotgun are the only 12-gauge shotguns currently authorized for firing 12-gauge nonlethal projectile munitions. The launching cup for the rubber ball grenade will attach to the Mossberg 500A2 Shotgun. The launch cup will not attach to the Mossberg 590 shotgun. Until the Joint Service Combat shotgun can effectively fire nonlethal 12-gauge rounds, the 12-gauge Mossberg 500A2 or 590 shotgun configuration meets the requirement of the NLW force. (Figure A-33)

Figure A-34. 12-Gauge M870 (modified) Version 2 (USCG)

42. 12-Gauge M870 (modified) Version 2 (USCG): The modified Remington 870 shotgun has a fixed synthetic stock and pistol grip. The pistol grip gives the shooter more control of the weapon and helps dissipate recoil from the shoulder. The synthetic materials used are more durable and lighter weight than the wood stocks used on the Standard M870. The barrel is fitted with a rifled choke tube for firing slugs (solid projectiles) to increase accuracy. It is also equipped with an Advanced Combat Optical Gunsight that was developed for use in "both-eyes-open" close quarters battle. The advantage is rapid target acquisition without the need to align front and rear sights. Batteries do not power the illumination sight; it contains tritium gas.

In the Coast Guard, a yellow "stock sock" will be used whenever the shotgun is loaded with nonlethal munitions. This removable nylon sleeve is used to identify the weapon as a nonlethal delivery system. (Figure A-34)

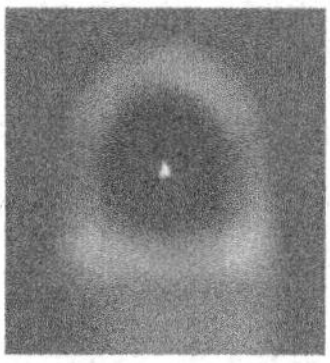

Figure A-35. Advanced Combat Optical Gunsight (ACOG) Reflex Sight (USCG)

43. Advanced Combat Optical Gunsight (ACOG) Reflex Sight (USCG): Modified shotguns are equipped with this site. (Figure A-34)

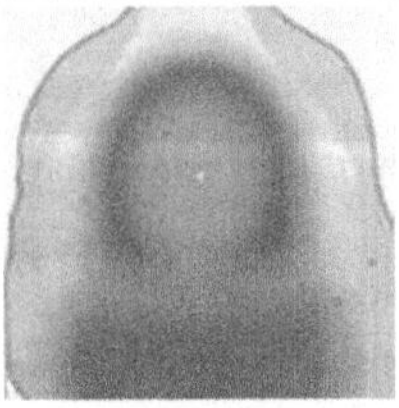

Figure A-36. ACOG Reflex Sight (USCG)

44. ACOG Reflex Sight (USCG): The dual-illuminated, Trijicon® Reflex sight was developed for the military for use in both-eyes-open close quarters battle (CQB). The advantage is quick target acquisition thus eliminating the need to align front and rear sights. Batteries do not power the illumination sight; it contains tritium gas. (Figure A-35)

Equipment Currently Being Fielded

1. 40-mm Crowd Dispersal Cartridge (CDC):

This direct fire, low-hazard, non-shrapnel-producing, multiple ball projectile provides friendly forces the capability to conduct crowd control. The CDC uses nonlethal, blunt trauma to stop, confuse, disorient, and/or temporarily incapacitate personnel and increase standoff distance between friendly forces and the hostile crowd. Primarily designed as an area effect munition, the CDC can be used on individuals. The M203 40-mm grenade launcher fires the CDC and contains 48, .48 caliber rubber balls. Effective range is 10 to 30 meters. (Figures A-37 and A-38)

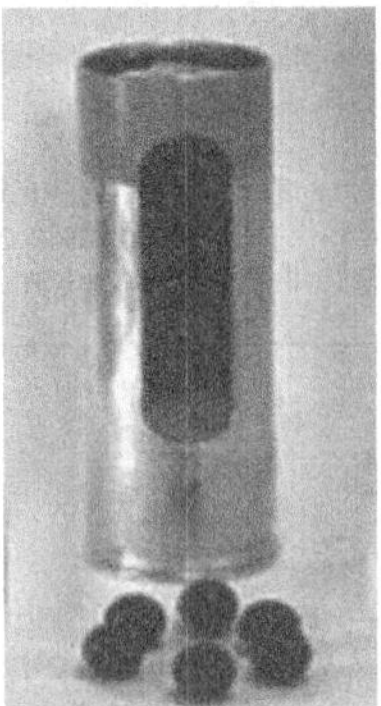

Figure A-37. Crowd Dispersal Cartridge

Figure A-38. Soldier Employing CDC with M203

2. Portable Vehicle Arresting Barrier (PVAB):

PVAB is a lightweight, portable, manually emplaced and recoverable barrier entrapment device to create, on command, a barrier that stops vehicles driven by known or suspected hostile forces with minimal damage to the vehicle or its occupants. It is used at dismount points, checkpoints, and roadblocks to prevent unauthorized access by wheeled vehicles into or out of areas (e.g. airfields, ports, command posts [CP]). Two individuals can easily emplace it, in approximately four hours. In stand-by mode, PVAB folds down into something similar to a speed bump (vehicles pass freely). In capture mode, the net springs up on two hydraulic pistons, to maximum height in less than a second. PVAB can be set up to span 12 to 24 feet and can stop a light truck (up to 7,500 pounds) at speeds up to 45 miles per hour, with a stopping distance of less than 200 feet PVAB can be containerized: 53.75 cubic feet (7 boxes, 4 large-3 small)—1,044 pounds on approximately one pallet. (Figures A-39 and A-40)

Figure A-39. PVAB in Action (Far View)

Figure A-40. PVAB in Action Near View

3. Modified Crowd Control Munition—Ground Emplaced (MCCM-GE):

The MCCM-GE (Figure A-41) is a nonlethal munition, housed in a casing similar to the current claymore mine, which will stop, confuse, disorient and/or temporarily incapacitate area targets/personnel at close range. The force can use it to provide crowd control by dispersing hostile crowds and force protection by integrating it into a layered perimeter defense. It delivers 600, .32 caliber rubber balls as well as a flash bang effect. The MCCM-GE gives the commander or small unit leader greater flexibility to apply nonlethal force as a first line of defense against hostile noncombatants. Effective range is 5 to 15 meters. (Figure A-42)

Figure A-41. Modified Crowd Control Munition-Ground Emplaced (MCCM-GE)

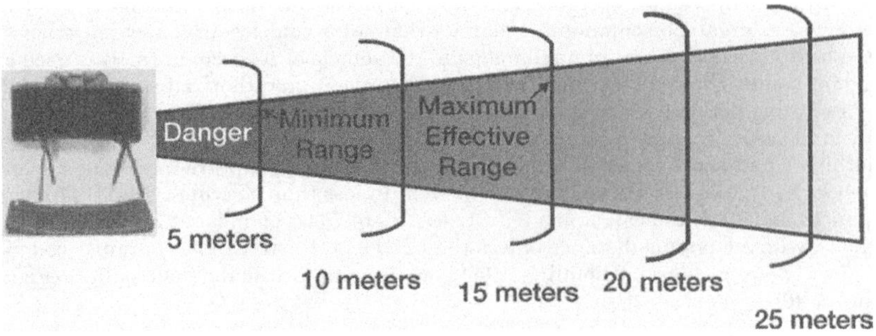

Figure A-42. MCCM-GE Range Chart

4. 66-mm Vehicle Mounted Nonlethal Grenade Launcher (VMNLGL):

The VMNLGL is a nonlethal grenade-launching device mounted on high mobility multipurpose wheeled vehicle (HMMWV). It is an indirect fire-support system that can deliver a 66-mm distraction device (flash-bang), CS, or blunt trauma (sting ball) payload. The distraction grenade uses a pyrotechnic charge for crowd control through audio and visual stimuli. The blunt trauma grenade uses a pyrotechnic charge to discharge 140, .32 caliber (cal.), rubber balls to achieve crowd control through audio, visual, and physical stimuli. The force uses VMNLGL to enhance the ability of friendly forces to conduct force protection without direct contact with hostile or threatening crowds. It enforces standoff distances and deters potential threats. It can disperse or subdue crowds, keep personnel from surrounding or climbing on patrol vehicles, slow or stop threatening personnel or vehicles by confusing, disorienting, and/or temporarily incapacitating individuals for capture. The grenade is fired from the grenade launchers traditionally found on tracked vehicles. Vehicle-launched grenades have three submunitions designed to ground burst. Effective range is 50 to 100 meters. An adjustable bracket allows engagement at ranges of 50 meters, 75 meters, and 100 meters. (Figure A-44)

Figure A-43. 66-mm Vehicle Mounted Nonlethal Grenade Launcher (VMNLGL)

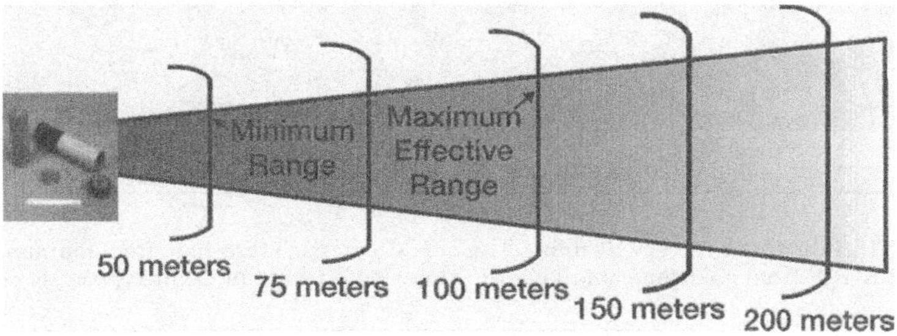

Figure A-44. VMNLGL Range Chart

5. Running Gear Entanglement System (RGES)

The Running Gear Entanglement System (RGES) is series of short, separate, looped polyethylene lines that are attached to a single, long polyethylene base line. This entanglement net fouls the propellers of planing hull small boats. The capability to manually deploy the net from a soft-sided bag has been fielded for USCG use in limited numbers. The USCG is in the process of incorporating three other variants of RGES into its inventory of nonlethal capabilities. These include aerial-delivered RGES, surface launched RGES to deploy from vessels, and static barrier RGES. The static barrier RGES has broad application into support homeland security, physical security and force protection missions. The surface launched RGES has direct application in the USCG's law enforcement mission wherever the requirement exists to stop a non-compliant vessel. Both of these systems are expected to be fielded in the fall/winter of CY02. The air launched version is currently still in the concept development phase and is not yet fully developed. (Figure A-45)

Figure A-45. Running Gear Entanglement System (RGES)

a. RGES Surface Delivery System: The USCG can use stern-mounted, manually deployed RGES from a cutter's small boat to stop slow-moving migrant-type craft. (Figure A-46)

(1) Uses COTS compressed air cylinder launcher.

(2) Removable, fix-mounted launcher on side of craft.

(3) Launch projects device ahead of target.

Figure A-46. RGES Surface Delivery System

b. RGES Barrier

(1) Provides static barrier around ship or other waterside asset.

(2) Provides a visible demarcation line, establishes zones, deters intrusions. (Figure A-47)

Figure A-47. RGES Barrier Employment

c. RGES Aerial Delivery System (ADS): RGES is aerial deployed from the ADS ahead of target aboard a H60 helicopter (Figure A-48). Device should float in the water ahead of target ensuring target runs over device. Device fouls propeller or rudder until target stops and awaits boarding.

(1) Uses COTS compressed air cylinder launcher

(2) Removable, fix-mounted launcher on helicopter

(3) Launch projects device ahead of target. (Figures A-49)

Figure A-48. RGES Aerial Delivery System (ADS)

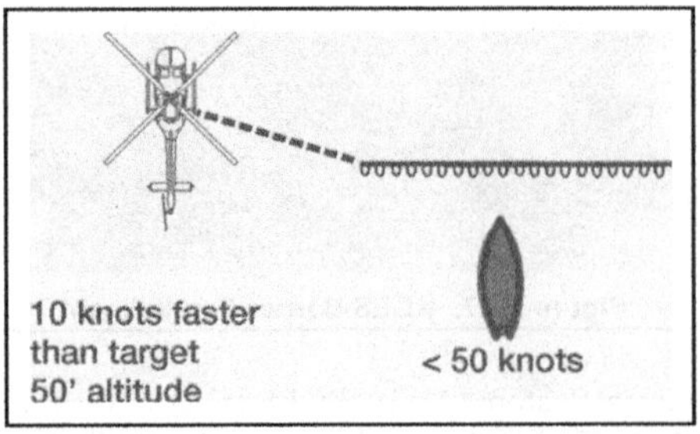

Figure A-49. RGES Aerial Delivery System (ADS)

6. Warp Cutter

a. Warp cutting is a nonlethal use of force, which has been safely and effectively employed by several countries to enforce fisheries laws in their respective exclusive economic zones (EEZ). The warp cutter is a device that closely resembles a grappling hook or small anchor. It has hardened steel blades attached to the crux of the grapnels. It is towed behind an enforcement vessel and across the warps of a non-compliant fishing vessel. As the cutting device engages the warps, it slices through them, separating the net from the fishing vessel.

b. The USCG uses the warp cutter to compel foreign vessels suspected of illegally fishing in the U.S. EEZ to comply with lawful orders to stop and be boarded. The USCG may cut the trawl warps of non-compliant vessels as a means of compelling them to stop for a boarding.

Figure A-50. Warp Cutter

Appendix B

Summary: Specialty Impact Munitions

12-Gauge Launch Cartridge

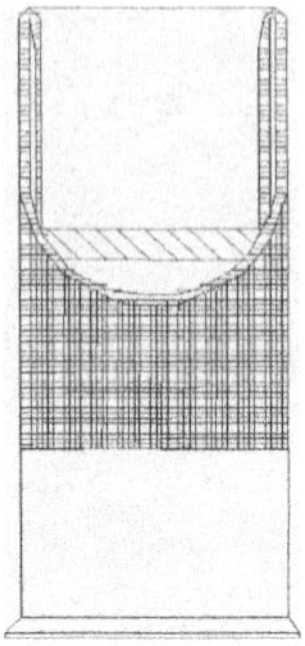

Using Services: USN, USMC
Status: Fielded 1998
DODIC: AA30
NSN: 1305-01-464-8389
Acquisition Activity: MARCORSYCOM—NSWC Crane, IN
Further Information:

Commercial: MARCORSYSCOM (703) 784-9395
NSWC Crane (812) 854-5801

Color....................................Clear case
Diameter...............................12-gauge
Projectile..............................N/A
Length...................................2.00 inches (51 mm)
Total Weight.........................13.6 grams
Propellant Weight.................46.3 grains (black powder)
Muzzle Velocity.....................100 feet/second (30.5 m/sec)—rubber ball grenade
Acceptance Accuracy..........200 feet (61 m) + 25 feet (7.6 m) at 30-degree launch angle using rubber ball grenade and 12-gauge launch cup

Projectile Specification: N/A

- Use 12-gauge launch round in combination with rubber ball grenade and 12-gauge launch cup to hurl a grenade a distance of 200 feet (61 m) + 25 feet (7.6 m) when launch angle is approximately 30 degrees.
- Chamber launch cartridge only after installing grenade in launch cup and pulling safety pin from grenade.

Figure B-1. 12-Gauge Launch Cartridge

12-Gauge Bean Bag Cartridge

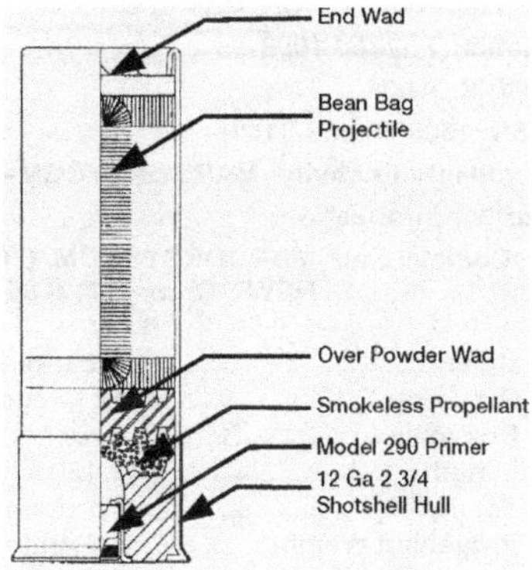

Using Services: USN, USMC
DODIC: AA29
NSN: 1305-01-454-0191
Status: Fielded 1998
Acquisition Activity: MARCORSYCOM—NSWC Crane, IN
Further Information:
Commercial: MARCORSYSCOM (703) 784-9395
NSWC Crane (812) 854-5801

Color....................................Clear or red case
Diameter................................12 gauge
Projectile...............................1 bean bag
Length...................................2.45 inches (62 mm)
Total Weight..........................49.9 grams
Propellant Weight.................3.4 grains
Muzzle Velocity.....................300 + 25 feet/second (91.4 + 7.62 m/sec)
Acceptance Accuracy..........2 feet (.6 m) X 6 feet (1.8 m) at 100 feet (30.3 m) with Mossberg 500A2

Shotgun Projectile Specification:
Shape...............................Square bag or tube sock
Material............................Nylon or cotton-polyester
Filler.................................#9 lead shot
Filler weight.....................44.4 grams (maximum)

- Fatal injuries are possible at employment distances of less than 20 feet (6 m)
- At distances of 20 feet (6 m) to 40 feet (12 m), target area should be center mass
- Training needs to reinforce that head shots are NOT ACCEPTABLE
- Do not skip fire this round! Round becomes unpredictable upon striking the ground
- Round is most effective against individually selected targets (point round)
- End wad has low light marker

Figure B-2. 12-Gauge Bean Bag Cartridge

12-Gauge Rubber Fin Stabilized Cartridge

Using Services: USN, USMC, USCG

Status: Fielded 1998

DODIC: AA31

NSN: 1305-01-454-0189

Acquisition Activity: MARCORSYCOM—NSWC Crane, IN

Further Information:

Commercial: MARCORSYSCOM (703) 784-9395
NSWC Crane (812) 854-5801

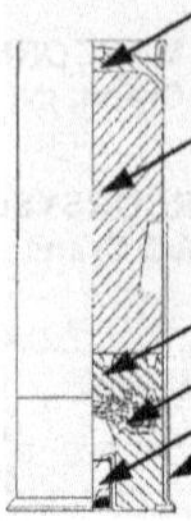

Color ....Clear or red case
Diameter ....12 gauge
Projectile ....1 rubber projectile
Length ....2.45 inches (62 mm)
Total Weigh ....9.1 grams
Propellant Weight..................4.8 grains
Muzzle Velocity......................530 + 65 feet/second
(161.5 + 19.8 m/sec)
Acceptance Accuracy..........2 feet (.6 m) X 6 feet (1.8 m) at 100 feet (30.3 m) with Mossberg 500 A2 shotgun
Projectile Specification: ......Caliber 12 gauge
Material..................................Rubber
Hardness ....Durometer 70-80 Shore "A" scale
Weight ....7.8 grams
Temperature Limits (demonstrated):
Upper limit...125° F
Lower limit...0° F

Shipping Data: Hazard classification/division and storage compatibility group...1.4S
DOT Class ....ORM-D
DOT Markings Cartridge......Small Arms
UNO serial number...............0012

- Fatal injuries are possible at employment distances of less than 20 feet (6 m); USCG: 15 yards
- At distances of 20 feet (6 m) to 40 feet (12 m), target area should be center mass; USCG: 15 to 30 yards
- Training needs to reinforce that head shots are NOT ACCEPTABLE
- Do not skip fire this round! Round becomes unpredictable upon striking the ground
- Round is most effective against individually selected targets (point round)
- End wad has low light marking

Figure B-3. 12-Gauge Rubber Fin Stabilized Cartridge

40-mm Foam Rubber Baton Cartridge

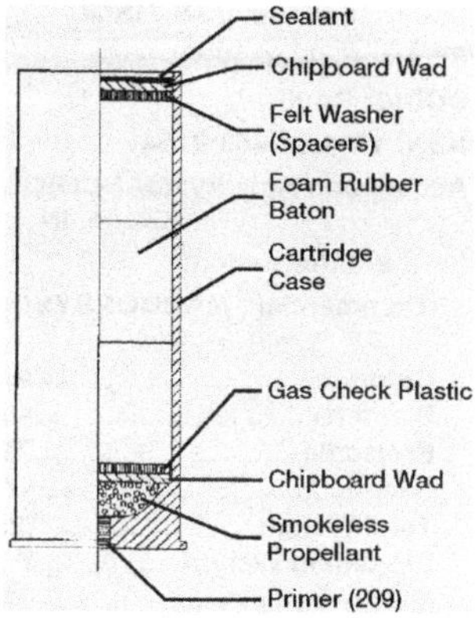

Using Services: USN, USMC
Status: Fielded 1998
DODIC: BA07
NSN: 1310-01-453-9168
Acquisition Activity: MARCORSYSCOM—NSWC Crane, IN
Further Information:
Commercial: MARCORSYSCOM (703) 784-9395
NSWC Crane (812) 854-5801

Color .... **Aluminum (case)**
Diameter .... **40 millimeter**
Projectile .... **3 foam batons**
Length .... **4.78 inches (121.4 mm)**
Total Weight .... **201 grams**
Propellant Weight .... **13.0 grains**
Muzzle Velocity .... **325 + 35 feet/second (99 + 10.6 m/sec)**
Acceptance Accuracy .... **6 feet (1.8 m) X 3 feet (.9 m) at 50 feet (15.2 m) with M203**
Projectile Specifications:
Number **3**
Material .... **Foam rubber**
Hardness .... **Durometer 25 + 10 on Shore "A" scale**
Weight .... **250 to 280 grains**

- **Fatal injuries are possible at employment distances of less than 20 feet (6 m)**
- **At distances of 20 feet (6 m) to 40 feet (12 m), target area should be center mass**
- **Training needs to reinforce that head shots are NOT ACCEPTABLE**
- **Round is most effective against point targets (individual personnel)**
- **End wad has low light marker**

Figure B-4. 40-mm Foam Rubber Baton Cartridge

40-mm Rubber Ball Cartridge

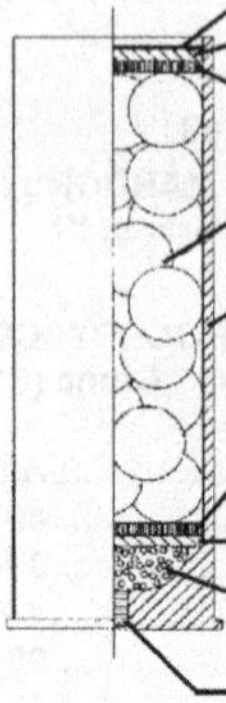

Using Services: USN, USMC
Status: Fielded 1998
DODIC: BA08
NSN: 1310-01-453-9154
Acquisition Activity: MARCORSYSCOM—NSWC Crane, IN
Further Information:
Commercial: MARCORSYSCOM (703) 784-9395

Color....................................Aluminum (case)
Diameter................................40 millimeter
Projectile...............................24 rubber balls
Length...................................4.78 inches (121.4 mm)
Total Weight...........................201 grams
Propellant Weight.................7.3 grains
Muzzle Velocity.....................325 + 35 feet/second (99 + 10.6 m/sec)
Acceptance Accuracy...........6 feet (1.8 m) x 9 feet (2.7 m) at 50 feet (15.2 m) with M203NSWC Crane (812) 854-5801

Projectile Specifications:
Caliber...................................60
Number..................................22 (minimum)
Material..................................Rubber
Hardness................................Durometer 60 + 10 on Shore "A" scale
Weight....................................2.2 grams each

- Fatal injuries are possible at employment distances of less than 20 feet (6 m)
- At distances of 20 feet (6 m) to 40 feet (12 m), target area should be center mass
- Training needs to reinforce that head shots are NOT ACCEPTABLE
- Round is most effective against area targets (multiple personnel)
- End wad has low light marker

Figure B-5. 40-mm Rubber Ball Cartridge

Rubber Ball Grenade

Using Services: USN, USMC

Status: Fielded 1998

DODIC: GG04

NSN: 1310-01-454-0132

Acquisition Activity: MARCORSYSCOM—NSWC Crane, IN

Further Information:

Commercial: MARCORSYSCOM (703) 784-9395
NSWC Crane (812) 854-5801

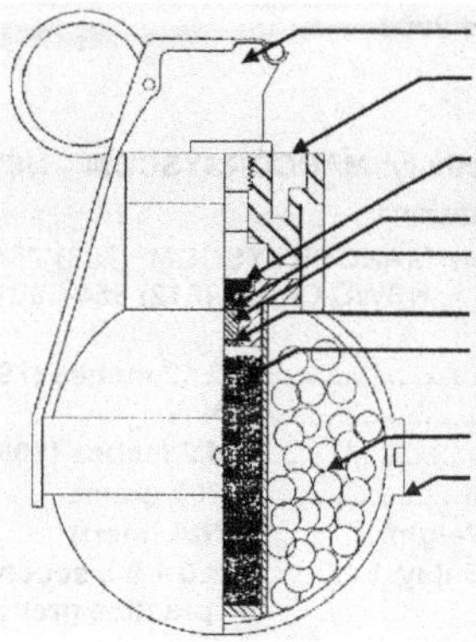

Color	Black with white lettering
Diameter	3.12 inches (79 mm)
Projectiles	100 rubber balls (minimum)
Height	5.25 inches (133.4 mm)
Total Weight	290 grams
Explosive Weight	59 grams
Total Time Delay	3.0 + 0.5 seconds

Employment Distance:

Hand Thrown	50 feet (15.2 m)
Shotgun Launched	200 feet (61 m)

Projectile Specifications:

Number	100 (minimum)
Material	Rubber
Hardness	Durometer 70 to 80 on Shore "A" scale
Diameter	0.25 inches (6.4 mm) (minimum)

- Fuze separates from the body 1.3 to 1.7 seconds after primer ignition.
- Second delay causes burster charge detonation 3.0 + 0.5 seconds average after primer ignition. Rubber balls are then dispersed in a 50-feet (15-m) circular pattern.
- 12-gauge shotgun with launch cup and launch cartridge can launch grenade 200 feet (61 m) + 25 feet (7.6 m)

Figure B-6. Rubber Ball Grenade

Practice Grenade Body

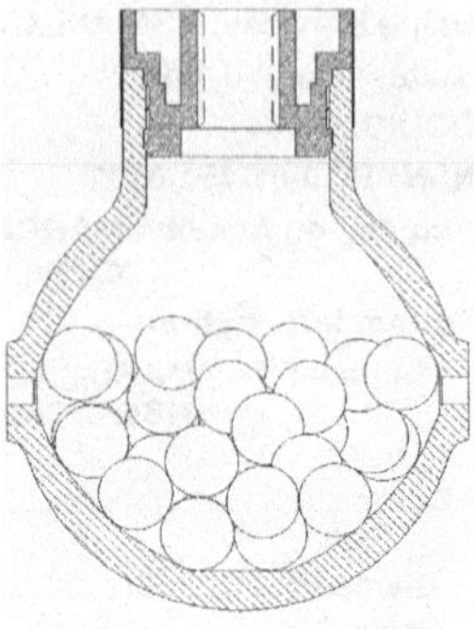

Using Services: USMC
Status: Fielded 2002
DODIC: GG05
NSN: Pending
Acquisition Activity: MARCORSYSCOM – NSWC Crane, IN
Further Information:
Commercial: MARCORSYSCOM (703) 784-9395
NSWC Crane (812) 854-5801

Diameter................................3.12 inches (79 mm)
Projectiles.............................N/A
Height....................................4.2 inches (106 mm)
Total Weight...........................208 grams
Explosive Weight.................N/A (inert)
Total Time Delay...................3.0 + 0.5 seconds when used with practice grenade fuze
Employment Distance:
Hand Thrown...................50 feet (15.2 m)
Shotgun Launched.........200 feet (61 m) using Mossberg launch cup and 12-gauge launch cartridge (AA30).

- Practice grenade body is reusable up to 25 times and is a component of a practice grenade assembly, consisting of the body and a practice grenade fuze.
- Practice grenade assembly accurately replicates the GG04 rubber ball grenade and is used for training purposes.

Figure B-7. Practice Grenade Body

Practice Grenade Fuze

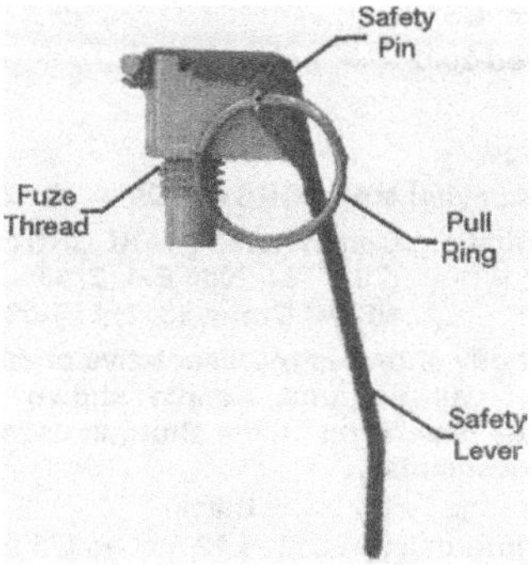

Using Services: USMC

Status: Fielded 2002

DODIC: Pending

NSN: Pending

Acquisition Activity: MARCORSYSCOM—NSWC Crane, IN

Further Information:

Commercial: MARCORSYSCOM
(703) 784-9395
NSWC Crane (812) 854-5801

Color .... Blue safety lever with brown stripe
Projectiles .... N/A
Height .... 4.25 inches (107 mm)
Total Weight............................ 82 grams
Explosive Weight..................... Not available
Total Time Delay...................... 3.0 + 0.5 seconds

Employment Distance when used with GG05 Practice Grenade Body:

Hand Thrown .... 50 feet (15.2 m)
Shotgun Launched........... 200 feet (61 m)—using Mossberg launch cup and 12-gauge launch cartridge (AA30).

- Employ practice grenade fuze with the practice grenade body to create a practice grenade assembly consisting of the body and a practice grenade fuze.
- Practice grenade assembly accurately replicates the GG04 rubber ball grenade and is used for training purposes.

Figure B-8. Practice Grenade Fuze

Nonlethal Grenade Launch Cup

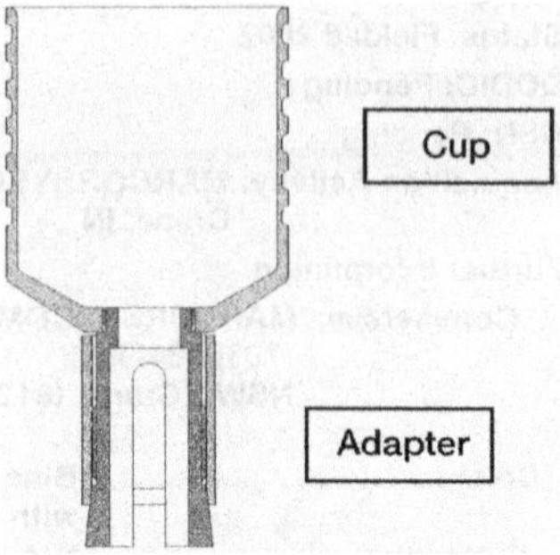

Cross Section View

Using Service: USMC

Status: Fielded 2002

DODIC: N/A

NSN: Pending

Acquisition Activity: MARCORSYSCOM—NSWC Crane, IN

Further Information: Commercial: MARCORSYSCOM
(703) 784-2006 Ext. 2733
NSWC Crane (812) 854-5801

NOTE: Cup body shown is representative of cup bodies used on all shotguns. Adapter shown is subject to change depending on the shotgun used and design improvements.

Color .... Black
Diameter (internal) .... 3.12 inches (79 mm)
Length (without adapter) ..... 4 inches (102 mm)
Weight .... 397 grams (maximum)

- Use launch cup in concert with AA30 launch cartridge and either the GG04 rubber ball grenade or GG05 practice grenade assembly
- Attach launch cup and remove from the shotgun without using tools
- Launch cup can withstand accidental detonation of rubber ball grenade without fragmenting

Figure B-9. Nonlethal Grenade Launch Cup

Diversionary Charge, MK 141, MOD 0

Using Services: USN, USMC
Status: Fielded 1997
DODIC: DWBS
NSN: 1375-01-387-5748
Acquisition Activity: MARCORSYSCOM—NSWC Crane, IN
Further Information:
Commercial: MARCORSYSCOM (703) 784-9395
NSWC Crane (812) 854-3934

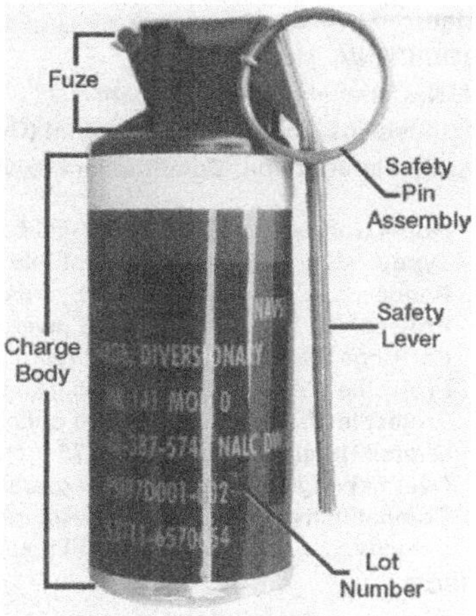

Color	Brown with white stripe
Diameter	1.77 inches (45 mm)
Projectile	N/A
Length	5.214 inches (132.4 mm)
Total Weight	99.8 grams
Net Explosive Weight	17.8 grains
Light Output	2 million candlepower for 4 milliseconds
Noise Output	176 to 185 decibels
Material	Molded and machined polyurethane
Time Delay:	
Initial	1.26 seconds (minimum)
Total	1.34 to 1.84 seconds

- Diversionary charge can cause serious injury if held at time of detonation
- Light and noise most effective in confined space to stun individuals
- Most often employed for hostage rescue

Figure B-10. Diversionary Charge, MK 141, MOD 0

Cartridge, 12-Gauge Rubber Sting Ball™

Using Services: USCG
Status: Fielded
DODIC: N/A Model No. 2552
NSN: Commercial Procurement
Acquisitions Activity: Commandant (G-OCU)
Further Information: Commandant (G-OCU) (202) 267-1522

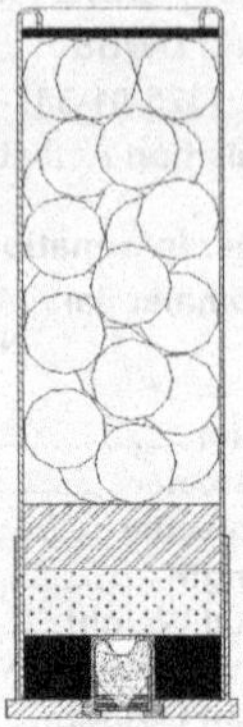

Model Number	31-caliber sting-ball
Type	Multiple projectile rubber pellets
Range	10 yards (9 m)
Projectiles	18 (average)
Cartridge Material	Plastic
Projectile Weight	.35 grams each
Projectile Material	.31-caliber rubber pellets
Overall Height	2.75 inches (70 mm)
Overall Weight	18 grams
Compatibility	12-gauge shotguns
Velocity	400 feet/second

USCG

- Fatal injuries are possible at employment distances of less than 5 yards.
- At distances of 20 feet (6 m) to 40 feet (12 m), target area should be center mass
- This round is most effective against area targets with a maximum effective range of 15 yards
- Training needs to reinforce that head shots are NOT ACCEPTABLE
- Round is most effective against area targets (multiple personnel)
- Stock Sock: a yellow "stock sock" will be used whenever the modified 12-gauge Remington 870 shotgun is loaded with nonlethal munitions. This removable nylon sleeve is used to identify the weapon as a nonlethal delivery system.

- All specifications are average and are subject to change without notice
- 2552SPEC

Figure B-11. Cartridge, 12-Gauge Rubber Sting Ball™

Cartridge, 12-Gauge Sting Ball™ High Velocity

Using Services: USCG
Status: Fielded
DODIC: N/A Model No. 2553
NSN: Commercial Procurement
Acquisitions Activity: Commandant (G-OCU)
Further Information: Commandant (G-OCU) (202) 267-1522

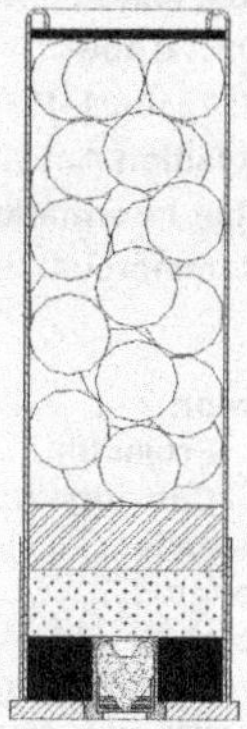

Model Number	31-caliber sting-ball
Type	Multiple projectile rubber pellets
Range	10 yards (9 m)
Projectiles	18 (average)
Cartridge Material	Plastic
Projectile Weight	.35 grams each
Projectile Material	.31-caliber rubber pellets
Overall Height	2.4 to 2.5 inches (61 – 63 mm)
Overall Weight	18 grams
Compatibility	12-gauge shotguns
Velocity	900 feet/second

USCG

- Fatal injuries are possible at employment distances of less than 5 yards.
- At distances of 20 feet (6 m) to 40 feet (12 m), target area should be center mass
- Maximum effective range is 15 yards
- Training needs to reinforce that head shots are NOT ACCEPTABLE
- Round is most effective against area targets (multiple personnel)
- Stock Sock: a yellow "stock sock" will be used whenever the modified 12-gauge Remington 870 shotgun is loaded with nonlethal munitions. This removable nylon sleeve is used to identify the weapon as a nonlethal delivery system.

- All specifications are average and are subject to change without notice.
- 2553SPEC

Figure B-12. Cartridge, 12-Gauge Sting Ball™ High Velocity

M1006, 40-mm Nonlethal Cartridge (Sponge Grenade)

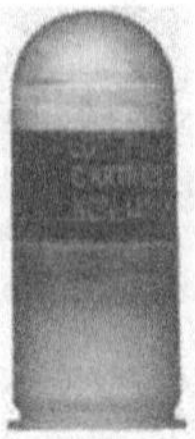

Using Services: U.S. Army

Status: Materiel Released April 2000

DODIC: BA06

NSN: 1310-01-452-1190

Acquisition Activity: Acquisition Center, Picatinny Arsenal, NJ

Further Information:

Commercial: Picatinny Arsenal NL Systems Integrator: (973) 724-6948
Product Director: (973) 724-6283

Color:
- Projectile .... Black, olive green (foam rubber)
- Cartridge Case .... Opaque

Diameter .... 40 millimeter
Length .... 5.272 inches (13.4 cm) (maximum)
Total Weight.......................... 68 grams (0.15 lbs)
Propellant Weight................. 2 grains
Muzzle Velocity..................... 265 + 20 feet/second (81 + 6 m/sec)
Acceptance Accuracy .... >85 % accuracy at 30 meters (E-silhouette) and >60 % accuracy at 50 meters (E-silhouette) when fired from an M16A2/M203 or a M4/M203
Engagement Range.............. 10 meters to 50 meters
Projectile Specification:
- Shape .... 40-mm bullet shaped foam rubber
- Material............................... Foam rubber
- Weight .... 28.5 grams (.06 lbs)

- Fatal injuries are possible at employment distances of less than 32.8 feet (10 m)
- At distances of 10 meters to 50 meters, target area should be center mass
- Training needs to reinforce that head shots are NOT ACCEPTABLE
- Do not skip fire this round! Round becomes unpredictable upon striking the ground
- Round is most effective against individually selected targets (point round)

Figure B-13. M1006, 40-mm Nonlethal Cartridge (Sponge Grenade)

M1012, 12-Gauge Shotgun: Nonlethal, Point Control Cartridge

Using Services: U.S. Army
Status: Type Classified, June 2000
DODAC: AA51
NSN: 1305-01-470-2405
Acquisition Activity: Acquisition Center, Picatinny Arsenal, NJ
Further Information:
Commercial: Picatinny Arsenal NL Systems Integrator: (973) 724-6948
Product Director: (973) 724-6086

Color:
Cartridge Case: ....Clear
Diameter................................12 gauge
Length ....2.45 inches (62.2 mm)
Total Weight...........................500 grains
Propellant Weight.................6.6 grains smokeless (Red Dot)
Muzzle Velocity.....................500 feet/second (152.4 m/sec)
Acceptance Accuracy..........80 % accuracy at 10 and 20 meters (E-silhouette) fired out of Mossberg 500, Mossberg 590, and Winchester 1200
Engagement range...............10 to 20 meters
Projectile Specification:
Shape ....Fin stabilized 'bomblet' shape
Material..............................Rubber
Weight ....0.2 ounces (5.8 gm)
Hardness........................... 75 Durometer "A" Scale

- Fatal injuries are possible at employment distances of less than 33 feet (10 m)
- At distances of 10 to 20 meters, target area should be center mass
- Training needs to reinforce that head shots are NOT ACCEPTABLE
- Round is most effective against individually selected targets (point round)

Figure B-14. M1012, 12-Gauge Shotgun: Nonlethal, Point Control Cartridge

M1013, 12-Gauge Shotgun: Nonlethal, Crowd Dispersal Cartridge

Using Services: U.S. Army

Status: Type Classified June 2000

DODAC: AA52

NSN: 1305-01-470-2139

Acquisition Activity: Acquisition Center, Picatinny Arsenal, NJ

Further Information:

Commercial: Picatinny Arsenal NL Systems Integrator: (973) 724-6948
Product Director :(973) 724-6086

Color:
- Projectile .... Black rubber
- Cartridge Case .... Clear

Diameter .... 12 gauge
Projectile .... 18 rubber balls
Length .... 2.37 inches (60.2 mm)
Total Weight .... 500 grains (32.4 gm)
Propellant Weight .... 14 grains smokeless (Red Dot)
Muzzle Velocity .... 900 feet/second (274 m/sec)
Acceptance Accuracy .... 80 % accuracy for a 2-by-3-meter target at 10 and 20 meters fired out of Mossberg 500, Mossberg 590, and Winchester 1200
Engagement range .... 10 to 20 meters
Projectile Specification:
- Shape .... 32-gauge (0.32 in) ball
- Material .... PVC rubber compound
- Weight .... 0.01 ounces (0.25 gm)
- Hardness .... 75 Durometer "A" Scale

- Fatal injuries are possible at employment distances of less than 32.8 feet (10 m)
- At distances of 10 to 20 meters, target area should be center mass
- Training needs to reinforce that head shots are NOT ACCEPTABLE

Figure B-15. M1013, 12-Gauge Shotgun: Nonlethal, Crowd Dispersal Cartridge

M1029, 40-mm Nonlethal Crowd Dispersal Grenade

Using Services: U.S. Army
Status: Type Classified May 2001
DODIC: BA13
NSN: 1310-01-475-0628
Acquisition Activity: Acquisition Center, Picatinny Arsenal, NJ
Further Information:

Commercial: Picatinny Arsenal NL Systems Integrator: (973) 724-6948
Product Director: (973) 724-6283

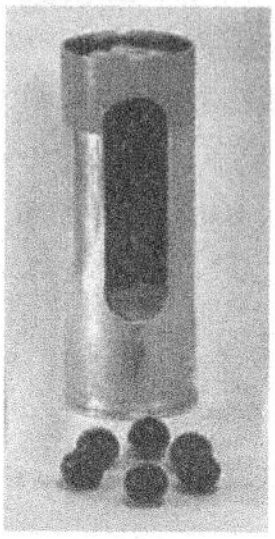

Color:
- Projectile .... Black rubber
- Cartridge Case .... Aluminum

Diameter .... 40 millimeter
Projectile .... 48 rubber balls
Length .... 4.8 inches (12.2 cm)
Total Weight........................... 202.22 grams (7.0 oz)
Propellant Weight.................. 8.5 grains smokeless
Muzzle Velocity...................... 450 feet/second
Acceptance Accuracy.......... > 90 % hit accuracy 4 out of 5 (shoulder to shoulder) E-silhouettes target at 30 meters and when fired from an M16A2/M203 or a M4/M203

Projectile Specification:
- Shape .... 48-gauge ball (.48 in diameter)
- Material............................. Rubber
- Weight .... 1.3 grams (.046 oz)
- Hardness........................... Durometer 60, scale "A"

- Fatal injuries are possible at employment distances of less than 32.8 feet (10 m)
- At distances of 10 to 30 meters, target area should be center mass of group of individuals
- Round is most effective when used to disperse groups of individuals

Figure B-16. M1029, 40-mm Nonlethal Crowd Dispersal Grenade

L96A1*, Grenade, Discharger: Antiriot, Irritant, CS

Using Services: U.S. Army, USMC

Status: Materiel Release, 4 January 2002

DODIC: FZ14

NSN: 1330-01-459-4018

Acquisition Activity: Acquisition Center, Picatinny Arsenal, NJ

Further Information:

Commercial: Picatinny Arsenal NL Systems Integrator: (973) 724-6948
Project Officer: (410) 436-2139

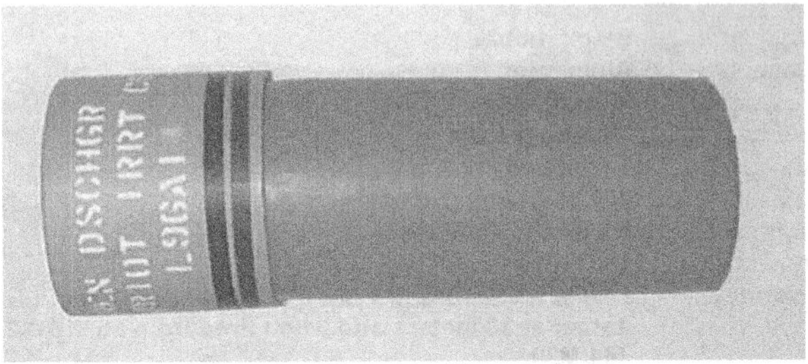

Color ....Light gray with red and brown bands
Diameter ....66 millimeters
Projectile ....23 canisters filled with CS compound
Length ....185 millimeters (7.28 in)
Total Weight..........................568 grams (1.25 lbs)
Propellant Weight................. 3.8 grams (.008 lbs)
Muzzle Velocity..................... 35.8 meters/second (117.5 ft/s)
Acceptance Accuracy..........65 to 95 meters
Deployment Range...............65 to 95 meters
Projectile Specification:

NOTE: It is not accurate to refer to the submunitions of the L96A1 grenade as projectiles. The antiriot material (CS) is released from the submunitions as they burn on the ground.

Shape...............................Cylindrical
Material............................Aluminum
Weight..............................11.3 grams (.025 lbs)

* The L97A1 Grenade is the training grenade for the L96A1 Grenade.

Figure B-17. L96A1, Grenade, Discharger: Antiriot, Irritant, CS

L97A1, Grenade, Discharger: Antiriot, Practice

Using Services: U.S. Army, USMC

Status: Material Released, 4 January 02

DODIC: FZ15

NSN: 1330-01-459-4032

Acquisition Activity: Acquisition Center,
Picatinny Arsenal, NJ

Further Information:

Commercial: Picatinny Arsenal NL Systems
Integrator: (973) 724-6948
Project Officer:
(410) 436-2139)

Color ....Blue with brown and green bands
Diameter ....66 millimeters
Projectile ....23 canisters filled with Cinnamic Acid (CA)
Length ....185 millimeters (7.28 in)
Total Weight..........................568 grams (1.25 lbs)
Propellant Weight.................3.8 grams (.008 lbs)
Muzzle Velocity.....................35.8 meters/second (117.5 ft/s)
Acceptance Accuracy..........65 to 95 meters
Deployment Range...............65 to 95 meters
Projectile Specification:

NOTE: It is not accurate to refer to the submunitions of the L97A1 grenade as projectiles. The antiriot, practice material (CA) is released from the submunitions as they burn on the ground.

Shape..............................Cylindrical
Material ....Aluminum
Weight.............................11.3 gram (.025 lbs)

- CA smoke can cause irritation to the eyes, mucous membranes, and skin. Personnel entering a CA cloud must wear appropriate face protection such as an M17- or M40-series protective mask.
- The burning grenade canisters could start a fire in dry grasses. Keep fire extinguisher available.
- During firing training, personnel within 16 meters must wear single hearing protection. This includes personnel within vehicle.
- Do not fire grenades when personnel or equipment are within 150 meters covering a 90-degree arc around a firing discharger.
- Do not fire grenades at the 50- or 75-meter position on the M315 adjustable aiming bracket. Fire only at the 100-meter setting.

Figure B-18. L97A1, Grenade, Discharger: Antiriot, Practice

Grenade, Launcher: Nonlethal, Distraction, M98

Using Services: U.S. Army, USMC

Status: Type Classified, 17 October 2001

DODIC: FZ16

NSN: 1330-01-484-7773

Acquisition Activity: Acquisition Center, Picatinny Arsenal, NJ

Further Information: Commercial: Picatinny Arsenal NL Systems Integrator: (973) 724-6948 Project Officer: (410) 436-2139

Color ....Green and black with brown band
Diameter ....66 millimeters
Projectile ....3 bursting submunitions, with pyrotechnic charges for audio and visual stimuli
Length ....25.25 centimeters (9.94 in)
Total Weight...........................725.7 grams (1.6 lbs)
Propellant Weight.................5.0 grams (.011 lbs)
Muzzle Velocity......................48.9 meters/seconds (160.4 ft/s)
Acceptance Accuracy..........100 to 120 meters
Deployment Radius..............10 to 15 meters
Projectile Specification:

NOTE: It is not accurate to refer to the submunitons of the M98 grenade as projectiles. These submunitions are ground bursting and the payload of inert ballast is just for flight stabilization and to deliver the audio/visual stimuli to the required range.

Filler ....Inert Ballast

- There is a remote chance of starting a fire. Keep fire extinguisher available.
- Personnel within 17 meters of a firing discharger must wear single hearing protection. This includes personnel within vehicle.
- Avoid looking directly at a bursting grenade; it could result in temporary loss of vision, lasting 2 to 5 minutes.
- Personnel within 35 meters of an operating launcher or 0.5 meters of a bursting canister should wear safety or ballistic-type eye protection, military clothing with long sleeves and elevated shirt collar, and helmet headgear.
- Do not handle or load M98s within 50 meters of a frequency transmitter emitting at 175 MHz, 448 MHz, 503 MHz and 1430 MHz.
- Do Not fire grenades when personnel or equipment are within 160 meters covering a 90-degree arc around a firing discharger.

Figure B-19. Grenade, Launcher: Nonlethal, Distraction, M98

M99, Grenade, Launcher: Nonlethal, Blunt Trauma

Using Services: U.S. Army, USMC

Status: Type Classified—17 October 2001

DODIC: FZ17

NSN: 1330-01-484-7775

Acquisition Activity: Acquisition Center, Picatinny Arsenal, NJ

Further Information:

Commercial: Picatinny Arsenal NL Systems Integrator: (973) 724-6948
Project Officer: (410-436-2139)

Color .... Green and black with brown band
Diameter .... 66 millimeters
Projectile .... 3 bursting submunitions, each with 140 balls
Length .... 25.25 centimeters (9.94 in)
Total Weight........................... 725.7 grams (1.6 lbs)
Propellant Weight.................. 5.0 grams (.011 lbs)
Muzzle Velocity...................... 48.9 meters/second (160.4 ft/s)
Acceptance Accuracy.......... 100 to 120 meters
Deployment Radius.............. 5 to 10 meters
Projectile Specification:

NOTE: It is not accurate to refer to the submunitions of the M99 grenade as projectiles. These submunitions are ground bursting and the payload of inert ballast is just for flight stabilization and to deliver the audio/visual stimuli to the required range.

Shape............................... 32-caliber (5/16 in) ball
Material .... PVC
Weight.............................. 0.40 grams (.014 oz)
Hardness .... 90 Durometer

- There is a remote chance of starting a fire. Keep fire extinguisher available.
- Personnel within 17 meters of a firing discharger must wear single hearing protection. This includes personnel within vehicle.
- Avoid looking directly at a bursting grenade; it could result in temporary loss of vision, lasting 2 to 5 minutes
- Personnel within 35 meters of an operating launcher or 0.5 meters of a bursting canister should wear safety or ballistic-type eye protection, military clothing with long sleeves and elevated shirt collar, and helmet headgear.
- Do not handle or load M98s within 50 meters of a frequency transmitter emitting at 175 MHz, 448 MHz, 503 MHz and 1430 MHz.
- DO NOT fire grenades when personnel or equipment are within 160 meters covering a 90-degree arc around a firing discharger.
- To minimize risk of eye injury, use the M99 primarily against crowds of adults.

Figure B-20. M99, Grenade, Launcher: Nonlethal, Blunt Trauma

Appendix C

Interservice Nonlethal Individual Weapons Instructor Course

1. Introduction

This appendix provides an overview of the Interservice Nonlethal Individual Weapons Instructor Course (INIWIC). The INIWIC is a "train-the-trainer" course taught in 15 subcourses and divided into four sections. Section I provides INIWIC descriptive data that includes the purpose, scope, peacetime and mobilization training requirements, student prerequisites, and equipment/ammunition requirements necessary to successfully complete the course. Section II lists the total training days and hours required per academic subcourse. Section III gives a detailed description of each subcourse and terminal learning objectives that each student must meet to complete the course. Section IV is a listing of 15 annexes that comprise the INIWIC Program of Instruction (POI). Access the entire INIWIC, with history, updates, and lesson plans at the following Internet address:
http://www.wood.army.mil/mcdet/niwic/niwic.htm.

2. Course Descriptive Data

a. Course Title: Interservice Nonlethal Individual Weapons Instructor Course

b. Location: Marine Corps Detachment, Fort Leonard Wood, Missouri

c. Marine Corps service school code: A16H5A3

d. Other service course number: USAF—L5AZA3P071019

e. Military assistance program articles and service list number: P166810

f. Purpose: To certify selected DOD/DOT service persons as nonlethal individual weapons instructors.

g. Scope: This is a train-the-trainer course designed for E-4s and above. The objective of this course is to prepare students to train individuals and groups in the utilization and employment of NLW and to serve as the unit NLW subject matter expert. This course is designed for all occupational specialties.

h. Length: 10 training days

i. Curriculum Breakdown: 100 total hours.

(1) 2.00 Demonstration.

(2) 24.00 Demonstration/practical exercise.

(3) 22.50 Lecture.

(4) 5.00 Lecture/demonstration.

(5) 4.00 Live-fire exercise.

(6) 26.50 Practical application.

(7) 6.00 Performance exam.

(8) 4.00 Written exam.

(9) 6.00 Administrative time.

j. Maximum Class Capacity: 30.

k. Optimum Class Capacity: 25.

l. Minimum Class Capacity: 15.

m. Class Frequency: 12.

n. Prerequisites: Students must be selected by commanders to be nonlethal weapons instructors and be an E-4 or above.

o. MOS Received: None.

p. Quota Control: Formal Schools Branch, Training and Education Command, Quantico, Virginia.

q. Funding: Service or command funded.

r. Reporting Instructions: All students report with orders to the Fort Leonard Wood, Missouri, Billeting Office, Bldg # 470 no later than 2400 on the day prior to class convening. Government quarters are available; however, messing is not.

s. Instructor Staffing: Two USMC instructors and two U.S. Army instructors.

t. School Overhead: Course OIC and chief instructor

Training Support: Training support is provided by the USMC for all common training support items. Service-specific training items are provided by the respective service.

Table C-1. Required Equipment For INIWIC

Item	*Quantity*
Personnel Protectors:	
Non-Ballistic Riot Face Shield	40
Non-Ballistic Full Length Riot Shield	40
Non-Ballistic Shin Guards	40
Ballistic Riot Face Shield	20
Ballistic Full Length Riot Shield	20
Ballistic Shin Guards	20
Kevlar Helmets	40
Flak Jacket/Body Armor	40
Personnel Effectors:	
Disposable Handcuffs (Flex Cuffs)	300
12-Gauge Area (rubber ball)	1400
12-Gauge Beanbag	1400
12-Gauge Fin Stabilized	1400
12-Gauge Flash-bang	1400
12-Gauge OO Buck	xxxx
12-Gauge 7 1/2 shot	xxxx
12-Gauge Launching Cup Cartridge	1400
40-mm Area (rubber ball)	1400
40-mm Foam Baton (3 foam sub-munitions)	1400
40-mm Sponge Grenade	1400
MK141 Diversionary Grenade	600

Table C-1. Required Equipment For INIWIC

Item	*Quantity*
M84 Reloadable Diversionary Grenade	600
Sting Ball Hand Thrown or Shot launched Grenade	600
Mossberg 500A2 Shotgun Launch Cup	15
M203 Grenade Launcher	15
Mossberg 500 3" Shotgun	15
36" Riot Oak Baton w/Belt Ring	40
Monadnock® 24-36" Expandable Baton w/holster	40
M36 RCA Dispenser	40
Mk-4 Personal OC Dispenser	40
Mk-9 Team OC Dispenser	40
Mk-46 High Capacity High Output OC Dispenser	20
MK-46 Hardware Kit (Refill Kit)	10
M33A1 Squad Riot Control Agent Dispenser	10
Modular Crowd Control Munitions Ground Emplaced (MCCM-GE)	20
66-mm Vehicle Launched Nonlethal Grenade (VLNLG)	3
66-mm NL munitions	30
Mission Enhancers:	
Portable Bullhorn	5
40-mm Carrying Pouch	40
Diversionary Rubber Ball Grenade Pouch	40
12-Gauge Butt Cuff carriers	40
12-Gauge Utility 25-Round Pouch	40
Mk-4 Individual Carrying Pouch	40
Portable Public Address System w/Microphone	3
12-Volt Hand-Held Spotlight	5
Night Hunter High Intensity Light System w/Kit	5
12-Gauge High Intensity Light Set	5
Caltrops	5
Road Side Spike Strip	5
Portable Vehicle Arrest Barrier (PVAB)	3
Training Devices:	
Impact Training Suit	6
Striking Bags	15
Training Batons (24 in)	40
Training Batons (36 in)	40
Practice Grenade Body	40
Practice Grenade Fuse	540
• OPFOR Training Munitions (foreign objects)	xxxx
• Target Requirements (full or partial torso)	40
MK-4 Personal OC Dispenser (training)	300
MK-9 Team OC Dispenser (training)	300
Training Disposable Handcuffs (flex cuffs)	100
12-Gauge Dummy Rounds	100

Table C-2. INIWIC Class V Requirements

Type	*#Per Student*	*# Demo Rounds*	*Total Per Class*	*Total Per Year*
12-Gauge Area (rubber ball)	5	5	155	1400
12-Gauge Beanbag	5	5	155	1400
12-Gauge Fin Stabilized	5	5	155	1400
40-mm Area (rubber ball)	5	5	155	1400
40-mm Foam Baton	5	5	155	1400
Sting Ball Hand Thrown or Shotgun Launched Grenades	2	2	62	600
MK141 Diversion Devices	2	2	62	600

3. Training Days Summary

This course consists of 10 training days. The curriculum is broken down into three categories.

Table C-3. INIWIC Curriculum

NLW Tactics	***Hours***
Force Continuum/ROE	2.00
Crowd Dynamics/Control/Formations	14.00
Communication Skills	2.00
MOOTW	2.00
Tactical Considerations	2.00
	22.00
NLW Systems	***Hours***
Oleoresin Capsicum	4.00
Empty Hand Control Techniques	19.00
Riot Baton Techniques	8.00
Nonlethal Munitions and Equipment	8.00
Portable Vehicle Barrier	5.00
66-mm Grenade Launcher	2.00
Modular Crowd Control Munition	2.00
	48.00
Nonlethal Weapons Instructor	***Hours***
Techniques of Military Instruction	10.00
NLW Scenario Exercise	10.00
	20.00
Subtotal =	**90.00**
Examinations =	***4.00***
Administration =	***6.00***
Total Hours =	***100.00***

4. Course Body

The course body consists of 15 subcourses. A quick synopsis of each subcourse is included below.

a. **Instructor Development**. This subcourse teaches and reinforces basic military instructor methodology. It identifies characteristics of good instructors and four ways they can improve their teaching ability. Additionally, it illustrates rules of conducting military instruction as well as the steps in planning a lesson and teaching classes.

b. **Force Continuum**. This subcourse introduces the five levels of resistance as well as identifying the levels of the force continuum. It illustrates influencing the use of force decisions and the definitions relating to deadly force. Lastly, it identifies the three preconditions for deadly force and seven authorized occasions for use of deadly force.

c. **Crowd Dynamics/Crowd Control**. This subcourse outlines the differences between crowds, mobs, and riots, and teaches the student basic crowd control techniques to apply in various situations. Upon completion, the student will be able to instruct others in both the behavioral aspects of a crowd during a civil disturbance and also various techniques of controlling a civil disturbance. It will identify factors that affect both individual and crowd behavior and it identifies all crowd control equipment needed, or essential, for mission accomplishment.

d. **Empty Hand Control Techniques**. This subcourse identifies safety factors to consider when instructing empty hand control techniques by illustrating the four Ms (mind, mass, movement, muscle) of self-defense, psychomotor skills, and phases of psychomotor skills. It illustrates levels of training and identifies common pressure points of the human body. It identifies hinges, appropriate angles of approach, and recommended patterns of movement. Instructors will learn blocking techniques, striking skills, control holds, escort positions, and restraint devices.

e. **Baton Techniques**. This subcourse identifies riot baton nomenclature and safety standards when conducting riot baton training. It identifies methods of teaching psychomotor skills and the phases of teaching psychomotor skills. It highlights baton-striking techniques, blocking techniques, and retention techniques.

f. **Military Operations Other Than War (MOOTW)**. This subcourse defines operations other than war and the difference between MOOTW and declarations of war. It helps identify the primacy of political objectives and the strategic aspects of MOOTW. It illustrates the range of operations and the principles of operations, as well as the types of military operations other than war.

g. **Rules of Engagement (ROE)**. This subcourse defines ROE and specific terms that apply to these rules. It also defines the purpose of ROE and provides the student an understanding of policies underlying the need for ROE. It identifies three sources for ROE and the two parts of ROE that will not change. It explains how ROE relate to using and applying NLW.

h. **Oleoresin Capsicum (OC) Training**. This subcourse explains how the three levels of OC are created, as well as the nomenclatures for all aerosol canisters. It highlights the different spray patterns, methods of delivery, and cautions concerning the traumatic hydraulic needle effect. It explains employment considerations and the three levels of OC contamination. Additionally, it explains the physical and mental effects of OC and trains students on different grips, draw methods, and proper stances when employing OC. It explains procedures for decontamination, recovery, and first aid.

i. **Types of Formations**. This subcourse explains the elements and individual responsibilities within a riot control formation. Additionally, it trains proper grip, balance, holding, and retention of the riot control shield. It highlights leadership requirements by teaching different commands and cadence for conducting riot control formations. It illustrates the different types of formations and the support elements of a riot control operation

j. **Tactical Considerations**. This subcourse addresses the preparations necessary for crowd control operations and the graduated response of a crowd control force. It identifies support and information required from host nation police and civil authorities. It describes the employment measures of military forces in crowd control and MOOTW environments. It highlights the actions, optional measures, and guidance for using NLW in crowd control and MOOTW environments.

k. **Nonlethal Munitions and Employment**. This subcourse identifies the types of nonlethal munitions and their intended purpose. It illustrates how nonlethal munitions fall within established "use of force" continuums. It identifies the types, characteristics, and ranges of all currently employed nonlethal munitions.

l. **Communications Skills**. This subcourse identifies the areas of nonverbal communication employed to determine an individual's state of mind. This, along with an understanding of the importance of personal space, gives the student the tools to resolve conflict with the amount of force as applied in the force continuum.

m. **Portable Vehicle Arrest Barrier**. This subcourse teaches the employment and maintenance of a force protection asset assigned to the nonlethal capability set. The PVAB is designed to stop a 7,500-pound vehicle going 45 mph. The PVAB protects gates and check points.

n. **Modular Crowd Control Munition**. This subcourse teaches the employment of the MCCM; the nonlethal version of the claymore mine. The MCCM is a force protection asset designed to help protect static defensive positions or check points.

o. **M-315 66-mm Grenade Launcher**. This subcourse teaches the installation, employment, and maintenance of the M-315. The M-315 increases the current maximum range of nonlethal sting ball munitions from approximately 50 meters to 100 meters. Additionally, this weapons platform can deploy flash bangs, smoke, and CS munitions.

Appendix D

Sample: Unit Training Schedules/Firing Tables

Nonlethal Individual Weapons Training Course Training Schedule: Unit Training Schedule Number 1.

Training Day #1					
Time	***Subject***	***Location***	***Uniform***	***Instr***	***Ref/Notes***
0800	Administration Time	Classroom	U1		
0830	Force Continuum	Classroom	U1		
1000	ROE	Classroom	U1		
1200	Chow	Mess Hall	U1		
1300	MOOTW	Classroom	U1		
1500	Communication Skills	Classroom	U1		
1700	Secure/COB				

Administration:

Uniforms:

U1: BDU/utility uniform
U2: BDU/utility uniform with LBE/harness and canteens
U3: BDU/utility uniform with LBE/harness, canteens, flak jacket, and helmet
U4: BDU/utility uniform with LBE/harness, canteens, flak jacket, helmet, and assigned weapon

Transportation:
Weapons:
Ammunition:
Required Equipment:
Screen
Video projector with central processing unit
Televisions
VCR
Miscellaneous:
All personnel are required to have a notebook, pencil, and pen.
Handouts prepared and passed out.

Training Day #2					
Time	*Subject*	*Location*	*Uniform*	*Instr*	*Ref/Notes*
0800	Nonlethal Equipment Brief	Classroom	U1		
0900	Pressure Point Control Techniques/ Open Hand Control Techniques	Classroom	U1		
1000	Pressure Point Control Techniques/ Open Hand Control Techniques (Practical Application)	PT Field	U2		
1200	Chow				
1300	Pressure Point Control Techniques/ Open Hand Control Techniques (Practical Application)	PT Field	U2		
1700	Debrief/Secure/COB				

Administration:

Uniforms:

U1: BDU/utility uniform

U2: BDU/utility uniform with LBE/harness and canteens

U3: BDU/utility uniform with LBE/harness, canteens, flak jacket, and helmet

U4: BDU/utility uniform with LBE/harness, canteens, flak jacket, helmet, and assigned weapon

Transportation:

Weapons:

Ammunition:

Required Equipment:

Screen

Video projector with central processing unit

Flexicuffs (one set of cuffs per two students)

Riot Training Bags (one bag per two students)

Miscellaneous:

All personnel are required to have a notebook, pencil, and pen.

Handouts prepared and passed out.

Medical personnel on station.

Training Day #3					
Time	*Subject*	*Location*	*Uniform*	*Instr*	*Ref/Notes*
0800	Pressure Point Control	Classroom	U1		
	Techniques/Open Hand Control Techniques	Classroom	U1		
0900	Pressure Point Control Techniques/Open Hand Control Techniques (Practical Application)	PT Field	U1		
1200	Chow				
1300	Pressure Point Control Techniques/Open Hand Control Techniques (Practical Application)	PT Field	U2		
1700	Debrief/Secure/COB				

Administration:

Uniforms:

U1: BDU/utility uniform
U2: BDU/utility uniform with LBE/harness and canteens
U3: BDU/utility uniform with LBE/harness, canteens, flak jacket, and helmet
U4: BDU/utility uniform with LBE/harness, canteens, flak jacket, helmet, and assigned weapon

Transportation:

Weapons:

Ammunition:

Required Equipment:

Screen
Video projector with central processing unit
Flex cuffs (one set of cuffs per 2 students)
Riot Training Bags (one bag per 2 students)

Miscellaneous:

All personnel are required to have a notebook, pencil, and pen.
Handouts prepared and passed out.
Medical personnel on station.

Training Day #4					
Time	***Subject***	***Location***	***Uniform***	***Instr***	***Ref/Notes***
0800	Impact Weapons	Classroom	U1		
0900	Impact Weapons (Practical Application)	PT Field	U2		
1200	Chow				
1300	Impact Weapons (Practical Application)	PT Field	U2		
1700	Debrief/Secure/COB				

Administration:

Uniforms:
U1: BDU/utility uniform
U2: BDU/utility uniform with LBE/harness and canteens
U3: BDU/utility uniform with LBE/harness, canteens, flak jacket, and helmet
U4: BDU/utility uniform with LBE/harness, canteens, flak jacket, helmet, and assigned weapon

Transportation:

Weapons:
Baton (one baton per 2 students)
Training baton (one baton per 2 students)

Ammunition:

Required Equipment:
Screen
Video projector with central processing unit
Flex cuffs (one set of cuffs per 2 students)
Riot training bags (one bag per 2 students)
Three (3) riot training suits

Miscellaneous:
All personnel are required to have a notebook, pencil, and pen.
Handouts prepared and passed out.
Medical personnel on station.

Training Day #5					
Time	*Subject*	*Location*	*Uniform*	*Instr*	*Ref/Notes*
0800	Impact Weapons (Practical Application)	PT Field	U2		
1000	Crowd Dynamics/Crowd Control	Classroom	U1		
1130	Chow				
1230	Crowd Dynamics/Crowd Control	Classroom	U1		
1330	Riot Control Formations	Classroom	U1		
1430	Oleoresin Capsicum (OC) Aerosol Training	Classroom	U1		
1630	Debrief/Secure/COB				

Administration:

Uniforms:
U1: BDU/utility uniform
U2: BDU/utility uniform with LBE/harness and canteens
U3: BDU/utility uniform with LBE/harness, canteens, flak jacket, and helmet
U4: BDU/utility uniform with LBE/harness, canteens, flak jacket, helmet, and assigned weapon

Transportation:

Weapons:
Training baton (one baton per 2 students)
Baton (one baton per 2 students)

Ammunition:
Inert OC for classroom instruction

Required Equipment:
Riot training bags (one bag per 2 students)
Screen
Video projector with central processing unit
Three (3) riot training suits

Miscellaneous:
All personnel are required to have a notebook, pencil, and pen.
Handouts prepared and passed out.

Training Day #6					
Time	***Subject***	***Location***	***Uniform***	***Instr***	***Ref/Notes***
0800	OC Aerosol Training (Practical Application)	PT Field	U2		
1200	Chow				
1300	OC Aerosol Training (Practical Application)	PT Field	U2		
TBD	Debrief/Secure/COB				

Administration:

Uniforms:
U1: BDU/utility uniform
U2: BDU/utility uniform with LBE/harness and canteens
U3: BDU/utility uniform with LBE/harness, canteens, flak jacket, and helmet
U4: BDU/utility uniform with LBE/harness, canteens, flak jacket, helmet, and assigned weapon

Transportation:

Weapons:

Ammunition:
MK 4.inert OC (one per student)
MK 4 individual can OC (2 cans)

Required Equipment:
Five (5) large trash cans with liners
Two (2) water hoses with running water

Miscellaneous:
Medical personnel on station.

Training Day #7					
Time	***Subject***	***Location***	***Uniform***	***Instr***	***Ref/Notes***
0730	Movement to Range		U4		
0800	Nonlethal Munitions and Employment	Range Classroom	U1		
1030	Chow	Range			
1100	Safety Brief/Range Brief	Range	U4		
1130	Nonlethal Munitions (Live Fire)	Range	U4		
TBD	Evening Chow				
TBD	Safety Brief/Range Brief	Range	U4		
1800	Night Exercise Sun Set +30)	Range	U4		
TBD	Debrief/Clean Up	Range	U1		
TBD	Movement to Rear				
TBD	Secure/ COB				

Administration:

Uniforms:

U1: BDU/utility uniform
U2: BDU/utility uniform with LBE/harness and canteens
U3: BDU/utility uniform with LBE/harness, canteens, flak jacket, and helmet
U4: BDU/utility uniform with LBE/harness, canteens, flak jacket, helmet, and assigned weapon

Transportation: Yes

Weapons:

Assigned Weapons

Ammunition:

12-gauge rubber ball
12-gauge fin stabilized
12-gauge bean bag
12-gauge launching cup cartridge
40-mm foam baton
40-mm rubber ball
40-mm sponge grenade
Stingball grenades

Required Equipment:

Eight (8) 12-Gauge stinger launcher cup
Fifteen (15) targets
Eye and hearing protection
Fifteen (15) full-length riot shields
All students require munitions pouches, shin guards, face shields
Two (2) MRE per student
Video projector with central processing unit
Three (3) high-intensity light
Thirty (30) individual tactical lights

Miscellaneous:

Range safety officer.
Medical personnel on station.
Secondary means of communication (radios).
Conduct operational risk assessment.

Training Day #8					
Time	***Subject***	***Location***	***Uniform***	***Instr***	***Ref/Notes***
0700	Movement to Training Area #1		U2		
0730	Portable Vehicle Arresting Barrier (PVAB)	TA#1	U1		
0900	PVAB (Practical Application)	TA#1	U2		
1200	Chow				
1300	PVAB (Practical Application)	TA#1	U2		
1500	Debrief/Movement to Rear				
1530	Preventive Maintenance of PVAB				
1600	Secure/COB				

Administration:

Uniforms:
U1: BDU/utility uniform
U2: BDU/utility uniform with LBE/harness and canteens
U3: BDU/utility uniform with LBE/harness, canteens, flak jacket, and helmet
U4: BDU/utility uniform with LBE/harness, canteens, flak jacket, helmet, and assigned weapon

Transportation: Yes

Weapons:

Ammunition:

Required Equipment:
One (1) portable vehicle-arresting barrier (PVAB)
Video projector with central processing unit

Miscellaneous:
(1) MRE per student.

Training Day #9					
Time	***Subject***	***Location***	***Uniform***	***Instr***	***Ref/Notes***
0730	Movement to Range		U4		
0800	Modular Crowd Control Munition-Ground Emplaced (MCCM-GE)	Range Classroom	U1		
0900	Safety Brief/Range Brief	Range	U4		
0930	MCCM-GE (Practical Application	Range	U4		
1030	Chow				
1130	Vehicle Mounted Nonlethal Grenade Launcher (VMNLGL)	Range Classroom	U4		
1300	VMNGL	Range	U4		
1730	Debrief/Movement to Rear				
1800	Secure/COB				

Administration:

Uniforms:
U1: BDU/utility uniform
U2: BDU/utility uniform with LBE/harness and canteens
U3: BDU/utility uniform with LBE/harness, canteens, flak jacket, and helmet
U4: BDU/utility uniform with LBE/harness, canteens, flak jacket, helmet, and assigned weapon

Transportation: Yes

Weapons:

Vehicle Mounted Nonlethal Grenade Launcher (VMNLGL)

Ammunition:

(2) MCCM-GE, (2) 66-mm Riot Control Training Grenade (L97E1)

Required Equipment:
Eye and ear protection
Video projector with central processing unit

Miscellaneous:
Range safety officer.
Medical personnel on station.
Secondary means of communication.
Two (2) MRE per student.

Training Day #10

Time	Subject	Location	Uniform	Instr	Ref/Notes
0700	Movement Down Range				U4
0730	Tactical Considerations	MOUT Classroom		U1	
0930	Safety Brief/Range Brief	MOUT Facility		U4	
1000	Tactical Exercise* (Practical Application)	MOUT Facility		U4	
1200	Chow				
1300	Tactical Exercise (Practical Application)	MOUT Facility		U4	
1700	Debrief/Movement to Rear				
1730	Weapons Maintenance				
1800	Secure/COB				

Administration:

Uniforms:
U1: BDU/utility uniform
U2: BDU/utility uniform with LBE/harness and canteens
U3: BDU/utility uniform with LBE/harness, canteens, flak jacket, and helmet
U4: BDU/utility uniform with LBE/harness, canteens, flak jacket, helmet, and assigned weapon

Transportation: Yes

Weapons:
Assigned weapon
Inert OC individual (one per student)
Training batons (one per student)
Fifty (50) practice grenades

Ammunition:

Required Equipment:
Three (3) riot training suit
All students require munitions pouches, flex cuffs, shin guards, face shields
Video projector with central processing unit
Fifteen (15) full length riot shields

Miscellaneous:
Range safety officer.
Medical personnel present.
Secondary means of communication.
(2) MRE's per student.

UNITED STATES MARINE CORPS
Marine Corps Detachment
Fort Leonard Wood, MO 65473

1.	LESSON DESIGNATOR:	Interservice Nonlethal Individual Weapons Instructor Course (INWIC). 13A (INIWIC.13A)
2.	LESSON TITLE:	Scenario Practical Exercise
3.	SCHOOL/COURSE:	Nonlethal Individual Weapons Instructor Course
4.	STUDY ASSIGNMENT:	Review course curriculum
5.	ESTIMATED TIME:	8 hours
6.	INSTRUCTORS REQUIRED:	(1) primary, (1) alternate per ratio: 1:10
7.	INSTRUCTOR REFERENCES:	INIWIC POI
8.	DATE PREPARED:	September 2001
9.	PRESENTATION METHOD:	Practical Exercise
10.	TRAINING AIDS:	Warning Order and five paragraph operations order for exercise, rubber rifles/shotguns, batons, shields, Inert MK4 OC canister, bullhorn, radio
11.	STUDENT EQUIPMENT:	(1) 782 gear (1) Kevlar with face shield (1) Shin guards

5. Unit Training Schedule Number 2

Table D-1. NLW Training Course 5-Day Training Schedule

Day/date	*Time*	*Event*	*Location*	*Instructor*	*Notes*
Mon/07 Jan 02	0730-0745	Welcome Aboard	Bldg RR202	MSgt Dixon	
T-1	0745-0845	Force Continuum	SOTG	TBD	L
	0900-1015	MOAB	SOTG	TBD	L
	1030-1130	OC	SOTG	TBD	L
	1130-1300	Noon Meal	Mess Hall	Company GySgt	
	1300-1400	OC Practical Application	SOTG	TBD	PA
	1400-1700	OC Certification	SOTG	TBD	PA
Tues/08 Jan 02	0730-0830	ROE	SOTG	TBD	L
T-2	0845-0945	Crowd Control	SOTG	TBD	L
	1000-1130	MDTS	SOTG	TBD	L, PA
	1130-1300	Noon Meal	Mess Hall	Company GySgt	
	1300-1700	MDTS	SOTG	TBD	L, PA
Wed/09 Jan 02	0730-1130	Batons	SOTG	TBD	L, PA
T-3	1130-1300	Noon Meal	Mess Hall	Company GySgt	
	1300-1500	Batons	SOTG	TBD	L, PA
	1500-1700	Formations	SOTG	TBD	L
Thurs/10 Jan 02	0730-1130	Formations	SOTG	TBD	PA
T-4	1130-1300	Noon Meal	Mess Hall	Company GySgt	
	1300-1700	Nonlethal Munitions/shotguns/ M2O3/sustainment training	Square bay/M203	TBD	L, PA
Fri/11 Jan 02	0730-1130	Nonlethal Munitions/stinger ball grenades/shotgun/M2 03	Square Bay/M203	TBD	PA
T-5	1130-1300	Noon Meal	Mess Hall	Company GySgt	
	1300-1700	Nonlethal Munitions/stinger ball grenades/shotgun/M2 03	Nonlethal Range 1&2	TBD	PA

Notes:

L = Lecture
PA = Practical Application

MDTS = Manadnock Defensive Tactics System
MOAB = Management of Aggressive Behavior
SOTG = Special Operations Training Group (USMC training unit)

6. Sample Unit Training Schedule Number 3

Table D-2. NLW Training Course 10-Day Training Schedule

DAY/DATE	TIME	EVENT	LOCATION	INSTRUCTOR	NOTES
Mon/29 Oct 01	0730-0745	Welcome Aboard	Bldg RR-202	MSgt Dixon	
T-1	0745-0845	Force Continuum	PT Field	Sgt Whitworth	L
	0900-1015	MOAB	PT Field	GySgt German	L
	1030-1130	OC	PT Field	Sgt Hoskins	L
	1130-1300	Noon Meal	Mess Hall	Company Gunny	
	1300-1400	OC Practical Application	PT Field	Sgt Hoskins	PA
	1400-1700	OC Certification	PT Field	Sgt Hoskins	PA
Tues/30 Oct 01	0730-0830	ROE	PT Field	Cpl Fewox	L
T-2	0845-0945	Crowd Control	PT Field	SSgt Parrish	L
	1000-1130	MDTS	PT Field	Sgt Johnson	L, PA
	1130-1300	Noon Meal	Mess Hall	Company Gunny	
	1300-1700	MDTS	PT Field	Sgt Johnson	L, PA
Wed/31 Oct 01	0730-1130	Batons	PT Field	Sgt Johnson	L, PA
T-3	1130-1300	Noon Meal	Mess Hall	Company Gunny	
	1300-1500	Batons	PT Field	Sgt Hoskins	L, PA
	1500-1700	Formations	PT Field	GySgt German	L
Thur/1 Nov 01	0730-1130	Formations	PT Field	Sgt Whitworth	PA
T-4	1130-1300	Noon Meal	Mess Hall	Company Gunny	
	1300-1700	Nonlethal Munitions	Nonlethal Range 1&2	GySgt German	L, PA
Fri/2 Nov 01	0730-1130	Nonlethal Munitions	Nonlethal Range 1&2	SSgt Parrish	PA
T-5	1130-1300	Noon Meal	Mess Hall	Company Gunny	
	1300-1700	Nonlethal Munitions	Nonlethal Range 1&2	SSgt Parrish	PA
Mon/5 Nov 01	0730-1130	Nonlethal Grenades Formation Fire	Nonlethal Range 1&2	GySgt German	L, PA
T-6	1130-1300	Noon Meal	Mess Hall	Company Gunny	
	1130-1700	Formation Fire/Sustainment Training	Nonlethal Range 1&2	SSgt Blackler	PA
Tues/6 Nov 01	0730-0830	Introduction to MOOTW	PT Field	Maj Desy	L
T-7	0830-1130	Vehicle Checkpoints/Vehicle Searches	PT Field	Sgt Whitworth	L, PA
	1130-1300	Noon Meal	Mess Hall	Company Gunny	
	1300-1500	Vehicle	PT Field	SSgt Parrish	PA

Table D-2. NLW Training Course 10-Day Training Schedule

DAY/DATE	TIME	EVENT	LOCATION	INSTRUCTOR	NOTES
		Checkpoints/Vehicle Searches			
	1500-1700	Roadblocks/Personnel Searches	PT Field	SSgt Blackler	L, PA
Wed/7 Nov 01	0730-1130	Roadblock/Personnel Searches	PT Field	SSgt Blackler	PA
T-8	1130-1300	Noon Meal	Mess Hall	Company Gunny	
	1300	Operations Order	Nonlethal Classroom	MSgt Dixon	Company Staff
Thurs/8 Nov 01	0730-1000	Confirmation Brief	PT Field	Maj Desy	
T-9	1000-1200	Insert	Combat Town	SSgt Parrish	
	1200-0000	STX	Combat Town	SSgt Parrish	
Fri/9 Nov 01	0001-1200	STX	Combat Town	SSgt Parrish	
T-10	1200-1300	Cleanup/Debrief	Combat Town	Company Gunny	
	1300	Graduation	Combat Town	SOTG OIC	

Notes:
L = Lecture
PA = Practical Application
MOAB = Management of Aggressive Behavior
MDTS = Manadnock Defensive Tactics System
STX = Situational Training Exercise

Appendix E

NLW Capability Sets

This appendix provides the current NLW capability sets for the services. These sets continue to be upgraded and changed according to service needs and the availability of new technologies. The U.S. Army and USMC capability sets are standard for the service, while the USAF, USN, and USCG are still developing and fielding their sets.

1. U.S. Army NLW Capability Set

Table E-1 U.S. Army Nonlethal Capability Set (Version FY 2001)

	Name	*Quantity*	*Unit Cost*	*Item Total*
GSAPMDK5X250AFHC	Non-Ballistic Riot Face Shields	200	$42.50	$8,500.00
GSAA TIPMBS9	Non-Ballistic Riot Body Shields Paulson	150	$88.50	$13,275.00
GSAATIUSMCSG	Non-Ballistic Riot Shin Guards	150	$60.00	$9,000.00
PT702MTUSA	Ballistic Face Shields	10	$246.00	$2,460.00
PT2237	Ballistic Body Shields	10	$1,052.00	$10,520.00
GSASF6SL	Sure Fire Field Light	10	$185.82	$1,858.20
MON2800 w/BCS-1	36" Hickory Riot Baton with Grommet	200	$15.50	$3,100.00
EADBH	Eagle Duty Baton Holder OD	200	$5.50	$1,100.00
GSAATIA2M	Portable Bullhorn	10	$82.00	$820.00
AAPB35W1	Ground Mount Bullhorn	8	$815.00	$6,520.00
AAPBWH10001	Megavox Wireless MIC	8	$0.00	$0.00
AASS450	Megavox Stand	8	$0.00	$0.00
GSASF6P	Individual High Intensity Light	15	$42.74	$641.10
GSAATIEATAVLFP6POD	Eagle Sure Fire 6P Holster	15	$8.75	$131.25
GSASFP60	Individual High Intensity Light Accessory Kit	15	$12.53	$187.95
XENNCSUSA	High Intensity Light	8	$2,795.00	$22,360.00
NIKFLEXCUFF	Disposable Forearm/Ankle Cuffs (Single)	1000	$0.70	$700.00
NIKCUTTER	Flex Cuff Cutters	2	$9.00	$18.00
GSAATIEAMK4	MK-4 Pouch OD	200	$6.75	$1,350.00
GSADEF5149	Inert Individual Training (OC) Dispenser	200	$6.07	$1,214.00
GSASF622	12-Gauge Shotgun High Intensity Light	50	$139.65	$6,982.50
GSAATIEAUPMS	12-Gauge Utility (25-Round) Pouch	50	$23.65	$1,182.50
GSAATISSSCQTR	12-Gauge Buttcuff Carrier (6-Round)	50	$10.75	$537.50
GSAATISASDD60D	Diversionary/Sting Ball Grenade Pouch	25	$47.25	$1,181.25
GSAATIEA40MM6	40-mm Carrying Pouch	100	$13.65	$$1,365.00
GSAATIRSC3	Caltrop	250	$10.50	$2,625.00
GSAATIFS333	Riot Training Suit with Accessories	4	$638.50	$2,554.00
GSAMONUTB	Riot Training Bag	10	$80.00	$800.00
BNCCELLS	C-Cell Batteries	0	$1.00	$0.00
BNAAACELLS	AA Batteries	0	$0.75	$0.00
GSASFDL123A	3V Lithium Batteries	0	$2.85	$0.00
BN9V	9V Batteries	0	$2.00	$0.00
ATIPMI	Packaging Marking and Integration Fee			$20,300.00
	Hardware Tools		**Item Total**	**$125,683.25**
	U.S. Army NLW Capability Set	**5**	**$125,683.25**	**$628,416.25**

2. USMC Nonlethal Weapons Capability Set

a. USMC NLW Company-Size Set

Table E-2 U.S. Marine Corps FY02 Company-Size Nonlethal Capability Set

TAMCN C41202E	***NSN 9999-01-464-2285***	***SET A (Company Size)***		
NSN/Product ID	***Name***	***Quantity***	***Unit cost***	***Item Total***
GSAATINCSPMDK5	Non-Ballistic USMC Riot Face Shields (S/R)	200	$42.50	$8,500.00
GSAATIPMBS9	Non-Ballistic Riot Body Shields Paulson	60	$88.50	$5,310.00
GSAATIUSMCSG	Non-Ballistic Riot Shin Guards	200	$60.00	$12,000.00
GSAATI23/36	USMC 23"/36" Riot Baton (Expandable) (w/Holster)	200	$73.25	$14,650.00
GSAATITRAINING BATON	USMC Riot Baton (Practice) (Price TBD)	200	$20.00	$4,000.00
GSAATIA2M	Portable Bullhorn	12	$82.00	$984.00
GSAATIDC/W	USMC Disposable Handcuffs	400	$0.94	$376.00
GSAATIDC/T	USMC Disposable Handcuff Training Device	100	$0.94	$94.00
GSAATIEAMK4	MK-4 Pouch OD	200	$6.75	$1,350.00
GSADEF5049	MK-4 Individual Live (OC) Dispenser (Pepper Spray)	200	$7.29	$1,458.00
GSADEF5099	MK-9 Live Squad (OC) Dispenser (Pepper Spray)	40	$26.20	$1,048.00
GSATIEAMK9	MK-9 Pouch OD	40	$11.00	$440.00
GSAATIDEF5846L	MK-46 Live (OC) (Horizontal) (Pepper Spray)	12	$300.00	$3,600.00
GSAATIDEF1515	MK-46 Sling	12	$30.25	$363.00
GSAATIDEF5520	MK-46 Hardware Kit	4	$68.50	$274.00
GSAATIDEF5546	MK-46 Refill Unit (OC) Live (Pepper Spray)	4	$46.25	$185.00
GSADEF5149	MK-4 Inert Individual Training (OC) Dispenser	200	$6.07	$1,214.00
GSADEF5199	MK-9 Inert Squad Training (OC) Dispenser	40	$20.10	$804.00
GSAATIDEF5946L	MK-46 Inert Training (OC) (Horizontal) (w/Dye)	12	$240.00	$2,880.00
GSADEF5546I	MK-46 Refill Unit (OC) Inert Training (w/Dye)	4	$27.50	$110.00
GSAATIFS333	Riot Training Suit w/Accessories (F.I.S.T.)	12	$638.50	$7,662.00
GSAMONUTB	Riot Training Strike Bag	24	$80.00	$1,920.00
GSAATIRSC3	Caltrop	400	$10.50	$4,200.00
TBD	Stinger Road Side Spike Kit (15.5') (Price TBD)	12		
XENNLCSUSMC	High Intensity Light (Xenonics)	6	$2,975.00	$17,850.00
GSAATIEATAVLFP6POD	Eagle Sure Fire 6P Holster	60	$8.75	$525.00
GSASF6P	Individual High Intensity Light	60	$42.74	$2,564.40
GSASFP60	Individual High Intensity Light Accessory Kit	30	$12.53	$375.90
GSASFDL123A	3V Lithium Batteries	120	$2.85	$342.00
GSAATISSSCQTR	12-Gauge Buttstock Cuff	36	$10.75	$387.00
GSAATIEAUPMS	12-Gauge Utility (25-Round) Pouch	72	$23.65	$1,702.80
GSAATIEA40MM6	40-mm Carrying Pouch	36	$13.65	$491.40
GSAATISASDD60D	Diversionary/Sting Ball Grenade Pouch	60	$47.25	$2,835.00
CSI9512	12-Gauge Launching Cup (Availability and Price TBD)	36		
1005-01-371-5639/ 1005-01-251-8578	12-Gauge Mossberg 500A2 / 590 Shotgun	36	$250.00	$9,000.00
ATIPMI	Packaging Marking and Integration Fee			$11,578.00
	Hardware Totals		**Item Total**	**$121,073.50**

b. USMC NLW Platoon-Size Set

Table E-3 USMC FY02 NLW Capability Set—Platoon Size

TAMCN C41202E	***NSN 9999-01-464-2285***	***SET B (Platoon Size)***		
NSN/Product ID	***NAME***	***Quantity***	***Unit Cost***	***Item Total***
GSAATINCSPMDK5	**Non-Ballistic USMC Riot Face Shields (S/R)	50	$42.50	$2,125.00
GSAATIPMBS9	Non-Ballistic Riot Body Shields Paulson	15	$88.50	$1,327.50
GSAATIUSMCSG	Non-Ballistic Riot Shin Guards	50	$60.00	$3,000.00
GSAATI23/36	USMC 23"/36" Riot Baton (Expandable) (w/Holster)	50	$73.25	$3,662.50
GSAATITRAINING BATON	USMC Riot Baton (Practice) (Price TBD)	50	$20.00	$1,000.00
GSAATIA2M	Portable Bullhorn	3	$82.00	$246.00
GSAATIDC/W	USMC Disposable Handcuffs	100	$0.94	$94.00
GSAATIDC/T	USMC Disposable Handcuff Training Device	25	$0.94	$23.50
GSAATIEAMK4	MK-4 Pouch OD	50	$6.75	$337.50
GSADEF5049	MK-4 Individual Live (OC) Dispenser (Pepper Spray)	50	$7.29	$364.50
GSADEF5099	MK-9 Live Squad (OC) Dispenser (Pepper Spray)	10	$26.20	$262.00
GSATIEAMK9	MK-9 Pouch OD	10	$11.00	$110.00
GSAATIDEF5846L	MK-46 Live (OC) (Horizontal) (Pepper Spray)	3	$300.00	$900.00
GSAATIDEF1515	MK-46 Sling	3	$30.25	$90.75
GSAATIDEF5520	MK-46 Hardware Kit	2	$68.50	$137.00
GSAATIDEF5546	MK-46 Refill Unit (OC) Live (Pepper Spray)	2	$46.25	$92.50
GSADEF5149	MK-4 Inert Individual Training (OC) Dispenser	50	$6.07	$303.50
GSADEF5199	MK-9 Inert Squad Training (OC) Dispenser	10	$20.10	$201.00
GSAATIDEF5946L	MK-46 Inert Training (OC) (Horizontal) (w/Dye)	3	$240.00	$720.00
GSADEF5546I	MK-46 Refill Unit (OC) Inert Training (W/Dye)	2	$27.50	$55.00
GSAATIFS333	Riot Training Suit W/Accessories (F.I.S.T.)	3	$638.50	$1,915.50
GSAMONUTB	Riot Training Strike Bag	3	$80.00	$240.00
GSAATIRSC3	Caltrop	50	$10.50	$525.00
TBD	Stinger Road Side Spike Kit (15.5') (Price TBD)	3		
XENNLCSUSMC	High Intensity Light (Xenonics)	4	$2,975.00	$11,900.00
GSAATIEATAVLFP6POD	Eagle Sure Fire 6P Holster	50	$8.75	$437.50
GSASF6P	Individual High Intensity Light	50	$42.74	$2,137.00
GSASFP60	Individual High Intensity Light Accessory Kit	25	$12.53	$313.25
GSASFDL123A	3V Lithium Batteries	100	$2.85	$285.00
GSAATISSSCQTR	12-Gauge Buttstock Cuff	15	$10.75	$161.25
GSAATIEAUPMS	12-Gauge Utility (25 Round) Pouch	30	$23.65	$709.50
GSAATIEA40MM6	40-mm Carrying Pouch	50	$13.65	$682.50
GSAATISASDD60D	Diversionary/Sting Ball Grenade Pouch	50	$47.25	$2,362.50
CSI9512	12-Gauge Launching Cup (Availability and Price TBD)	15		
1005-01-371-5639/ 1005-01-251-8578	12-Gauge Mossberg 500A2 / 590 Shotgun	15	$250.00	$3,750.00
ATIPMI	Packaging Marking and Integration Fee			$3,000.00
	Hardware Totals		**Item Total**	**$43,471.25**

** Riot shield. Marine Corps security forces are currently evaluating to use in conjunction with each other as listed below. They are not currently deployed but are undergoing field evaluation.
(1) The six feet Armadillo Interlocking Shield for the hard baseline.
(2) The four feet.. Armadillo Interlocking Shield for the soft baseline.

3. USN NLW Capability Set

a. The USN is in the process of developing NLW capabilities and capability sets.* The sets will support the specific areas of operation, mission, and/or user.

b. The first version will address the force protection mission and will be based on other service-approved items and some longer-range commercial-off-the-shelf items currently being upgraded for military use. The mobile security forces, master at arms, and shipboard force protection units are the targeted users.

c. A notional capability set for the Mobile Security Forces is listed below.

Table E-4. Notional USN Capability Set

Name	*Nomenclature (Mfr. Representative)*	*Quantity*	*Unit Cost*	*Case Size*	*Cost Extended*
Fin Stabilize Round		1 Case	$5	100	$400
Stingshot		1 Case	$5	100	$400
Shotbag		1 Case	$5	100	$450
Flamethrower Round		1 Case	$8	100	$750
Tactical Blast Bursting Round		1 Case	$7	100	$650
Ultraflash Bursting Round		1 Case	$8	100	$750
Launchable Stun Munition	POP, Inc. T-444	1 Case	$60	24	$1,440
40-mm Sparkshot		1 Case	$35	24	$840
40-mm Stingshot		1 Case	$25	24	$600
40-mm Shotbag		1 Case	$25	24	$600
40-mm OC		1 Case	$28	24	$660
Stinger	POP, Inc. I-452	1 Case	$60	24	$1,440
Stun	POP, Inc. T-470	1 Case	$55	24	$1,320
Flash/Bang Multi	POP, Inc. D-451	1 Case	$65	24	$1,560
OC	POP, Inc. I-452	1 Case	$65	24	$1,560
12-Gauge Launching Blanks		200	$3		$600
Grenade Training Rounds		12	$65		$780
12-Gauge Training Rounds		24	$3		$72
40-mm Training Rounds		12	$20		$240
12-Gauge Shotguns	Standard Issue/Modified	12	$1,000		$12,000
12-Gauge Launching Adapter		6	$325		$1,950
Face Shields		15	$100		$1,500
Helmets		15	$200		$3,000

* In FY 01, the USN fielded an USMC capability set to Naval Air Station, Roosevelt Roads, Puerto Rico.

Table E-4. Notional USN Capability Set

Name	*Nomenclature (Mfr. Representative)*	*Quantity*	*Unit Cost*	*Case Size*	*Cost Extended*
Tactical Vest		15	$300		$4,500
Bullhorn		2	$100		$200
Medical Kit		2	$50		$100
Searchlights		2	$100		$200
Field Pack		15	$100		$1,500
Holsters/Belts		15	$50		$750
Ordnance Packaging, Handling, Storage and Transportation (PHS&T)					$1,500
Misc. (Certification, Training and Documentation)					$3,000
			Total Cost Basic Kit		**$45,312**

Table E-5. Notional USN Capability Set

Name	*Nomenclature (Mfr. Representative)*	*Quantity*	*Unit Cost*	*Case Size*	*Cost Extended*
Stingmore Mine (Std Body) with Rubber Ball Projectiles	POP, Inc. I-454	24	$350		$8,400
Stingmore Mine Stand		6	$100		$600
Electric Firing Boxes		8	$35		$280
Area Lighting					$1,000
Electric Generator					$500
Signs and Placards					$500
Personal NVGs and Opticals		6	$2,000		$12,000
Additional PHS&T					$2,000
Misc. (Certification, Training and Documentation)					$2,000
Total Cost Augmented Basic Kit					**$27,280**
Total Cost (Basic plus Augment)					$72,592
Portable Vehicle Arresting Barrier	U.S. Army	1	$40,000		$40,000
Running Gear Entanglement System (RGES)	NSWCDD	1000 ft system	$25,000		$25,000
Antimaterial Round (USCG)		24	$12		$288
Spike Strips		2	TBD		
Improved Sensor (s)					
RF			TBD		

Table E-5. Notional USN Capability Set

Name	Nomenclature (Mfr. Representative)	Quantity	Unit Cost	Case Size	Cost Extended
IR			TBD		
Acoustic			TBD		
Chemical Detection	SABRE 2000	1	$20,000		$20,000
Ordnance Packaging, Handling, Storage and Transportation (PHS&T)					$6,000
Misc. (Certification, Training and Documentation)					$2,000
Total Cost Vehicle Supplement					**$93,288**
Total Cost Basic, Augmented, and Vehicle Supplement					**$165,880**

4. USAF NLW Capability Set

Table E-6. USAF NLW Capability Set

NSN	Item Description	EA	Item $	Item Total
GSADEF5149	Inert Individual (OC) Training Dispenser	1	$6.07	$6.07
GSADEF5846H	Mid-Size Riot Control Dispenser (Chemical)	3	$318.75	$956.25
GSADEF5546	Riot Control Dispenser (OC) Agent Quarts	1	$46.25	$46.25
	12-Gauge Shotgun (Area CDC/Rubber Pellet)	36	$4.46	$160.56
	12-Gauge Shotgun (Point/Fin Stabilized)	36	$2.85	$102.60
CSI (USMC) GG04	Stun Hand (Sting Ball) Grenade	50	$35.00	$1,750.00
M84	Stun Hand Grenade/Flash-Bang Device	50	$78.00	$3,900.00
GSAPMDK5X/AF	Non-Ballistic Face Shields	13	$29.50	$383.50
GSAPMBS9	Non-Ballistic Full Length Riot Shield	13	$88.50	$1,150.50
GSAUSMCSG	Non-Ballistic Shin guards	13	$6.000	$780.00
GSAATI23/36	36" Riot Baton w/Holster	13	$73.25	$952.25
GSAA2	Portable Megaphone	1	$82.00	$82.00
EAAFCUSTOM	Tactical Vest Rough Estimate	3	$150.00	$450.00
HARG700	Riot Control Gloves	13	$52.20	$678.60
Champro	Protective Cups	13	$7.25	$94.25
	40-mm (Point) Sponge Round	54	$23.00	$1,242.00
	40-mm (Area) Crowd Dispersal Cartridge Grenade	54	$22.00	$1,188.00
GSAF333	Riot Baton Training Suit	1	$638.50	$638.50

Table E-6. USAF NLW Capability Set

NSN	*Item Description*	*EA*	*Item $*	*Item Total*
MONDC/W	Disposable Forearm/Ankle Cuffs (FLEX)	40	$0.94	$37.60
GSADEF5049/ EA	Individual (OC) Dispenser with Carrier	13	$7.29	$94.77
	12-Gauge Utility Pouch	2	$24.00	$48.00
	12-Gauge Buttstock Cuff	3	$11.00	$33.00
	40-mm Carrying Pouch	2	$14.00	$28.00
	Training Strike Bag	2	$80.00	$160.00
	Cost of Clip-Lok Boxes			$480.00
	Approximate freight shipped via commercial freight into CONUS			$1,500.00
		Total Estimate		**$16,942.70**
Proposal to add TASERS to USAF Capability Set after testing.				

5. Notional USCG NLW Capability Set

Table E-7. Notional USCG NLW Capability Set

Reference: (a) COMDTINST M16247.1B Maritime Law Enforcement Manual (MLEM) (b) Project Ericson Website	
Expandable Baton	
Restraint Devices	Hand Cuffs, Flex Cuffs, Leg Irons, Migrant Restraint Harness
Individual OC Spray	
Non-Ballistic Riot Face Shield	
Non-Ballistic Riot Body Shield	
Non-Ballistic Riot Shin Guard	
Running Gear Entanglement Systems	

The following information describes each of the components of the NLW capability set currently deployed by the USCG. (The USCG has not designated a standard set of nonlethal gear. The using unit determines the gear needs.)

6. Restraint Devices

a. Handcuffs are the preferred restraint devices when dealing with a small number of subjects. Handcuffs should be steel, chain-linked construction capable of being double locked, with nickel, blued, or stainless steel finish.

b. Using flexible handcuffs for the temporary restraint of subjects is also authorized. Since they do not require a key and are relatively easy to apply, flexible handcuffs are particularly useful and appropriate when dealing with large numbers of subjects (e.g. migrants).

c. Leg irons are authorized restraint devices and may be a preferred option for restraining subjects for extended periods of time.

d. Migrant restraint harnesses are authorized when dealing with large numbers of subjects.

7. Expandable Baton

The expandable baton is the only intermediate weapon authorized for USCG personnel to carry. It should have a steel shaft that is silver in color and corrosion resistant (e.g. chrome, nickel, stainless steel) and have no knurling or sharp edges. It should have a foam or similar, non-slip grip. It should be rigid when deployed and must be no less than 15 inches and no greater than 22 inches when expanded. Spring loaded or other trigger style batons are not authorized.

8. OC Pepper Spray

Only chemical irritants containing the active ingredient OC are authorized for USCG personnel to carry. The aerosol must contain no greater than 10 percent and no less than 5 percent OC concentration. It shall have a water-based carrier, be nonflammable, nontoxic and produce a straight stream delivery (not fog).

9. Helmet Riot Face Shield

Comparable to DOD type. Not currently deployed. Undergoing field evaluation.

10. Riot Shield

Two shields are currently being evaluated. A 24 x 48 inches clear lexan weighing 14 pounds and a smaller version measuring 20 x 36 inches. They are not currently deployed. Undergoing field evaluation.

REFERENCES

Joint

JP 1-02, *DOD Dictionary of Military and Associated Terms*, 12 Apr. 2001

JP 3-07, *Joint Doctrine for Military Operations Other Than War*, 16 Jun. 1995

JP 3-07.3, *JTTP for Peace Operations*, 12 Feb.1999

JP 3-57, *Doctrine for Joint Civil-Military Operations*, 8 Feb. 2001

JP 3-58, *Joint Doctrine for Military Deception*, 31 May 1996

JP 3-61, *Doctrine for Public Affairs in Joint Operations*, 14 May 1997

JP 4-0, *Joint Doctrine for Logistic Support of Joint Operations*, 6 Apr. 2000

Multi-Service

FM 3-100.12, MCRP 5-12.1C, NTTP 5-03.5, AFTTP(I) 3-2.34, *Risk Management* for *Multiservice Tactics, Techniques, and Procedures*, 15 Feb. 2001

FM 3-31.1, MCWP 3-36, *Army and Marine Corps Integration in Joint Operations*, 21 Nov. 2001

Army

AR 190-14, *Military Police Carrying of Firearms and Use of Force for Law Enforcement and Security Duties*, 12 Mar. 1993

AR 350-38, *Training Devices Policies and Management*, 15 Oct. 1993

AR 27-1, *Judge Advocate Legal Services*, 30 Sep. 1996

FM 3-0, *Operations*, 14 Jun. 2001

FM 3-05.30 (33-1), *Psychological Operations*, 19 Jun. 2000

FM 19-15, *Civil Disturbances*, 25 Nov. 1985

FM 27-10, *The Law of Land Warfare*, 18 Jul. 1956

FM 100-19, *Domestic Support Operations*, 1 Jul. 1993

TC 90-1, *Training for Urban Operations*, 1 Apr. 2002

TRADOC Pamphlet 525-5, *Force XXI Operations, A Concept for the Evolution of Full Dimensional Operations for the Strategic Army of the Twenty-First Century,* 1 Aug. 1994

TRADOC Pam 350-9, *TRADOC Training Devices for Army wide Use*, 30 Sep. 1993

TRADOC Regulation 350-70, *Systems Approach to Training Management, Processes, and Products*, 9 Mar. 1999.

TRADOC Regulation 385-2, chg.1, TRADOC Safety Program, 10 Oct. 2000.

Marine Corps

MCWP 3-33.4, *Domestic Support Operations*, Jul. 1993

I MEF FMF, After Action Report—UNITED SHIELD, 7 Jun. 1995

MCO 1510.112, *Individual Training Standards for Nonlethal Weapons*, 27 May 1998

MCO 5500.6F, *Arming of Security and Law Enforcement Personnel and the Use of Force*, 20 Jul. 1995

MCRP 5-12.1A, *The Law of Land Warfare.*

MCO 1630.4A, *Law Enforcement and Physical Security Activities*, 19 Jun. 1991.

MCO 1510.114, *Individual Training Standards (ITS) for Antiterrorism Force Protection (AT/FT) System*, 30 Jan. 1998.

MCWP 3-11.1, *Marine Rifle Company*, Feb. 1978

MCRP 3-11.1A, *Commander's Tactical Handbook*, 30 Jul. 1988. .

MCWP 3-11.2 *Marine Rifle Squad*

Marine Corps Gazette, *"Less-Lethal" Force in Operation UNITED SHIELD*, Sep. 1995.

Navy

Naval Doctrine Publication 1, Naval Warfare (U), 28 Mar. 1994

NWP 1-14M, *The Commander's Handbook on the Law of Naval Operations*, Oct. 1995

SECNAVINST 5500.29B, *Use of Deadly Force and the Carrying of Firearms by Personnel of the Department of the Navy in Conjunction with Law Enforcement Security Duties and Personal Protection*, 29 Sep. 1992

Air Force

AFI 31-201, *Security Police Standards and Procedures*, 4 Dec. 2001

AFI 31-207, *Arming and Use of Force by Air Force Personnel*, 1 Sep. 1999

AFMAN 31-201V6, *Civil Disturbance*, 17 May 2002

AFMAN 31-222, *Security Forces Use of Force Manual*, 1 Jun. 2001

Coast Guard

COMDTINST M16247.1B, Maritime Law Enforcement Manual, 9 Aug. 2001

Other

DODD 3000.3-R, *Policy for Nonlethal Weapons*, 9 Jul. 1996

DODD 3025.1, *Military Support to Civil Authorities (MSCA)*, 15 Jan. 1993

DODD 5100.77, *DOD Law of War Program*, 9 Dec. 1998

DODD 3025, *I-M Manual for Civil Emergencies*, 2 Jun. 1994

DODD 3025.12, *Military Assistance for Civil Disturbances*, 4 Feb. 1994

DODD 3025.13, *DOD Resources in Support of US Secret Service*, 13 Sep. 1985

DODI 5000.2, *Operation of the Defense Acquisition System*, 5 Apr. 2002

DODR 5000.1, *Defense Acquisition*, 23 Oct. 2000, Ch 1, 4 Jan. 2001

DOD Joint Concept for Nonlethal Weapons, 5 Jan. 1998

DODD 3025.12, *Military Assistance for Civil Authorities*, 18 Feb. 1997

DODD 3025.16, *Military Assistance for Civil Authorities Liaison Office (EPLO) Program,* 18 Dec. 2000

DODI 5000.2, *Operation of the Defense Acquisition System*, 5 Apr. 2002

Mission Need Statement (MNS), *Range of Nonlethal Force Capabilities (NLFC*), 10 Oct. 1996

A Guide for Effective Peacetime Employment of Reserve Components Units and Individuals, OASD RA (M&P), Dec. 2000

Defense and Intelligence Abbreviations and Acronyms, Nov. 1997

Emergency Management Assistance Compact—Guide Book and Operating Procedures, National Emergency Management Association, Aug. 2000

Inside the Navy, Marine Corps Nonlethal Weapons Experiment Underway at Camp Pendleton, CA, Vol. 9, No. 29, Jul. 22, 1996

Military Support To Civil Authorities—Interagency Disaster Resource Guide, National Interagency Civil-Military Institute, Jan. 2001

National Guard Bureau Military Support to Civil Authorities Lessons Learned, 6 Sep. 2000

The Robert T. Stafford Disaster Relief and Emergency Assistance Act, as amended 42 USC 1521, Jul.2001

Joint Universal Lessons Learned System, Operation UNITED SHIELD

Nonlethal Weapons Joint Mission Area Analysis/Joint Mission Needs Analysis, Dec. 2000

Interservice Nonlethal Individual Weapons Instructor Course (INIWIC) Program of Instruction, Oct.2000

Presidential Decision Directive PDD-39 U.S. Policy on Counterterrorism 21 Jun.1995

PDD-62 Combating Terrorism—Fact Sheet, 22 May 1998

The Federal Response Plan Notice of Change 7, Feb.1997; FEMA, Number 229 Change 11

Federal Response Plan, Apr 1999

GLOSSARY

PART I – ABBREVIATIONS AND ACRONYMS

A	
ACOG	advanced combat optical gunsight
ADS	aerial delivery system
AFDC	Air Force Doctrine Center
AFI	Air Force instruction
AFMAN	Air Force manual
ALSA	Air Land Sea Application
AMC	Army Materiel Command
AO	area of operation
AR	Army regulation
ATSC	Army Training Support Center
B	
bldg	building
BLT	battalion landing team
bn	battalion
C	
C4ISR	command, control, communications, computers, intelligence, surveillance, and reconnaissance
CA	civil affairs
cal	caliber
CANS	compressed air net system
CDC	crowd disposal cartridge
CMC	Commandant of the Marine Corps
COB	close of business
COC	combat operations center
COMDTPUB	Commandant, Coast Guard Publication
COMDTINST	Commandant, Coast Guard Instruction
CONUS	continental United States
COTS	commercial off-the-shelf
CP	command post
CQB	close quarters battle
CS	O-Chlorobenzyl-Malononitrile (riot control agent (tear gas))
CSSOC	combat service support operations center
cu ft	cubic feet

D

DA	Department of the Army
dB	decibel
DECON	decontamination
demo	demonstration
DM	designated marksman
DOD	Department of Defense
DODD	Department of Defense Directive
DODI	Department of Defense Instruction
DODIC	Department of Defense Identification Code
DODAC	Department of Defense Ammunition Code
DOT	Department of Transportation
DP	downed person
DPB	deployable pursuit boat
DTD	dated

E

EADBH	Eagle Duty Baton Holder
EEZ	exclusive economic zone
EOD	explosive ordnance disposal

F

FAST	Fleet Antiterrorism Security Team
FDP	food distribution point
FLEX	forearm/ankle cuff
FM	field manual
four M's	mind, mass, movement, muscle
FSE	fire support element
ft	feet

G

G1	Army or Marine Corps component manpower or personnel staff officer (Army division or higher staff, Marine Corps brigade or higher staff)
G2	Army or Marine Corps component intelligence staff officer (Army division or higher staff, Marine Corps brigade or higher staff)
G3	Army or Marine Corps component operations staff officer (Army division or higher staff, Marine Corps brigade or higher staff)

G4	Army or Marine Corps component logistics staff officer (Army division or higher staff, Marine Corps brigade or higher staff)
ga	gauge
GOTS	government off-the-shelf
H	
HMMWV	high mobility multipurpose wheeled vehicle
I	
IADS	Integrated Air Defense System
IAW	in accordance with
ID	identify/identification
IFF	identification, friend or foe
in.	inch
INIWIC	Interservice Nonlethal Individual Weapons Instructor Course
IPB	intelligence preparation of the battlespace
ITRO	Interservice Training Organization
J	
JMAA	Joint Mission Area Analysis
JTTP	joint tactics, techniques, and procedures
K	
KFOR	Kosovo forces
L	
LBE	load-bearing equipment
LOAC	Law of Armed Conflict
LP	listening post
LVOSS	light visual optical screening system
M	
M16	M16A2 rifle
M203	M203 Grenade Launcher
MARCORSYCOM	Marine Corps System Command
MCCDC	Marine Corps Combat Development Command
MCCM-GE	modified crowd control munition—ground emplaced
MCO	Marine Corps Order
MCPDS	Marine Corps Doctrinal Publication System
MCRP	Marine Corps reference publication

MCSF	Marine Corps security force
MCWP	Marine Corps warfighting publication
MDST	Monadnock Defensive Tactics System
MEDEVAC	medical evacuation
MEF	Marine expeditionary force
METT-T	mission, enemy, terrain and weather, troops and support available – time available
METT-TC	mission, enemy, terrain and weather, troops and support available – time available and civil considerations
MEU	Marine expeditionary unit
MIO	Maritime interception operation
MLEM	Maritime Law Enforcement Manual
MILSTRIP	military
mm	millimeter
MNS	mission need statement
MOAB	management of aggressive behavior
MOOTW	military operations other than war
MOS	military occupational specialty
MOUT	military operations in urban terrain
MP	military police
mph	miles per hour
MRE	meal, ready to eat
MTTP	multi-Service tactics, techniques, and procedures

N

NATO	North Atlantic Treaty Organization
NAVSOP	Navy standing operating procedures
NDC	Naval Doctrine Command
NLFC	nonlethal force capabilities
NLW	nonlethal weapon
NSN	national stock number
NSWC	Naval Surface Warfare Command
NWDC	Naval Warfare Development Command
NWP	Naval Warfare Publication

O

OC	Oleoresin Capsicum
OCAT	Oleoresin Capsicum aerosol training
OCONUS	outside the continental United States
OD	olive drab
OP	observation post

OPFOR	opposing force
P	
PA	public affairs
PAO	public affairs officer
PASGT	personnel armor system, ground troops
POI	program of instruction
PSYOP	psychological operations
PT	physical training
PVAB	portable vehicle arresting barrier
PVC	polyvinyl chloride
Q	
QRF	quick reaction force
R	
R&D	research and development
RCA	riot control agent
RGES	running gear entanglement system
ROE	rules of engagement
RRF	rapid reaction force
S	
S1	battalion or brigade manpower or personnel staff officer (Army; Marine Corps battalion or regiment)
S2	battalion or brigade intelligence staff officer (Army; Marine Corps battalion or regiment)
S3	battalion or brigade operations staff officer (Army; Marine Corps battalion or regiment)
S4	battalion or brigade logistics staff officer (Army; Marine Corps battalion or regiment)
SASO	stability and support operations
SB	shield bearer
sec	second
SECNAVINST	Secretary of the Navy instruction
SG	shotgun
SJA	staff judge advocate
SNCO	staff noncommissioned officer
SOP	standard operating procedure
SOTG	Special Operations Training Group
SROE	standing rules of engagement
SSC	small-scale contingency

T

TC	training circular
TF	task force
T/O	table of organization
TOC	tactical operations center
TRADOC	United States Army Training and Doctrine Command
TTP	tactics, techniques, and procedures

U

UN	United Nations
US	United States
USA	United States Army
USAF	United States Air Force
USAMPS	United States Army Military Police School
USCG	United States Coast Guard
USMC	United States Marine Corps
USN	United States Navy
USS	United States Ship

V

VCR	videocassette recorder
VLNL	vehicle launched nonlethal grenade
VMNLGL	vehicle mounted nonlethal grenade launcher

W

w/	with
WMD	weapons of mass destruction

PART II – TERMS AND DEFINITIONS

Air Force specialty code (AFSC). Primary job of a U.S. Air Force officer or enlisted person. Analogous to the Army and Marine Corps military occupational skill (MOS).

area of operation (AO). An operational area defined by the joint force commander for land and naval forces. Areas of operation do not typically encompass the entire operational area of the joint force commander, but should be large enough for component commanders to accomplish their missions and protect their forces. (JP 1-02)

battalion landing team (BLT). In an amphibious operation, an infantry battalion normally reinforced by necessary combat and service elements; the basic unit for planning an assault landing. (JP 1-02)

civil affairs. Designate active and Reserve Component forces and units organized, trained, and equipped specifically to conduct civil affairs activities and to support civil-military operations. (JP 1-02)

civil-military operations. The activities of a commander that establish, maintain, influence, or exploit relations between military forces, governmental and nongovernmental civilian organizations and authorities, and the civilian populace in a friendly, neutral, or hostile operational area in order to facilitate military operations, to consolidate and achieve operational U.S. objectives. Civil-military operations may include performance by military forces of activities and functions normally the responsibility of the local, regional, or national government. These activities may occur prior to, during, or subsequent to other military actions. They may also occur, if directed, in the absence of other military operations. Civil-military operations may be performed by designated civil affairs, by other military forces, or by a combination of civil affairs and other forces. (JP 1-02)

combat service support. The essential capabilities, functions, activities, and tasks necessary to sustain all elements of operating forces in theater at all levels of war. Within the national and theater logistics systems, it includes, but is not limited to, that support rendered by service forces in ensuring the aspects of supply, maintenance, transportation, health services, and other services required by aviation and ground combat troops to permit those units to accomplish their missions in combat. Combat service support encompasses those activities at all levels of war that produce sustainment to all operating forces on the battlefield. (JP 1-02)

command and control. The exercise of authority and direction by a properly designated commander over assigned and attached forces in the accomplishment of the mission. Command and control functions are performed through an arrangement of personnel, equipment, communications, facilities, and procedures employed by a commander in planning, directing, coordinating, and controlling forces and operations in the accomplishment of the mission. (JP 1-02)

commercial off the shelf (COTS). Describes items procured for military use through civilian contractors.

core capability. The set of functions that define a significant, stand-alone, operationally effective and suitable military capability such that, should no further development occur, the user will have received a significant capability.

designated marksman (DM). A rifleman stationed in an overwatch position above friendly forces, armed with a standard infantry weapon appropriate to his service that is mounted with a high-powered scope and loaded with lethal ammunition. The designated marksman uses his vantage point to identify agitators to forces below, eliminate lethal threats, and monitor the battlefield.

dispersion. A scattered pastern of hits around the mean point of impact of bombs and projectiles dropped or fired under identical conditions. In chemical and biological operations, the dissemination of agents in liquid or aerosol form. (JP 1-02)

fleet Marine force (FMF). A balanced force of combined arms comprising land, air, and service elements of the U.S. Marine Corps. A fleet Marine force is an integral part of a U.S. Fleet and has the status of a type command. (JP 1-02)

force protection. Security program designed to protect service members, civilian employees, family members, facilities, and equipment, in all locations and situations, accomplished through planned and integrated application of combating terrorism, physical security, operations security, and personal protective services, and supported by intelligence, counterintelligence, and other security programs. (JP 1-02)

government off-the-shelf (GOTS). Analogous to COTS, but describing items already procured by the U.S. government and ready for issue to military units.

host nation. A nation that receives the forces and/or supplies of allied nations, coalition partners, and/or NATO organizations to be located on, to operate in, or to transit through its territory. (JP 1-02)

intelligence preparation of the battlespace (IPB). An analytical methodology employed to reduce uncertainties concerning the enemy, environment, and terrain for all types of operations. Intelligence preparation of the battlespace builds an extensive database for each potential area in which a unit may be required to operate. The database is then analyzed in detail to determine the impact of the enemy, environment, and terrain on operations, and present it in graphic form. Intelligence preparation of the battlespace is a continuing process. Also called IPB. (JP 1-02)

interrogator-translator team (ITT). Marine Corps unit, usually working as a small detachment to a larger infantry unit, highly trained in interrogation techniques and fluent in adversary native language.

joint force commander. A general term applied to a combatant commander, subunified commander, or joint task force commander authorized to exercise combatant command (command authority) or operational control over a joint force. (JP 1-02)

logistics. The science of planning and carrying out the movement and maintenance of forces. In its most comprehensive sense, those aspects of military operations which deal with:

a. Design and development, acquisition, storage, movement, distribution, maintenance, evacuation, and disposition of materiel.

b. Movement, evacuation, and hospitalization of personnel.

c. Acquisition of construction, maintenance, operation, and disposition of facilities.

d. Acquisition of furnishing of services. (JP 1-02)

low-intensity conflict. Political-military confrontation between contending states or groups below conventional war and above the routine, peaceful competition among states. It frequently involves protracted struggles of competing principles and ideologies. Low intensity conflict ranges from subversion to the use of armed force. It is waged by a combination of means employing political, economic, informational, and military instruments. Low-intensity conflicts are often localized, generally in the Third World, but contain regional and global security implications.

NOTE: Also known as small wars or "brushfire wars," low-intensity conflicts are rebellions, insurrections, and the like characterized by adversaries dedicated to a cause but lacking command and control structure and modern equipment.

Marine expeditionary force (MEF). The largest of the Marine air-ground task force (MAGTF) and the Marine Corps principal warfighting organization, particularly for larger crisis or contingencies. It is task-organized around a permanent command element and normally contains one or more Marine divisions, Marine aircraft wings, and Marine force service support groups. The Marine expeditionary force is capable of missions across the range of military operations, including amphibious assault and sustained operations ashore in any environment. It can operate from a sea base, a land base, or both. Also called MEF. See also aviation combat element; combat service support element; command element; Marine air-ground task force; Marine expeditionary force (forward); Marine expeditionary unit; special purpose Marine air-ground task force; task force. (JP 1-02)

maritime interception operation (MIO). Means by which an exclusion zone at sea is enforced. Involves the visit of merchant ships, by armed boarding parties if necessary, bound to, through, or out of a defined area, to examine each ship's papers and cargo and to search for evidence of contraband. The provisions must be established by a sanctioning body and must be applied to ships of all nationalities.

military operations other than war (MOOTW). Operations that encompass the use of military capabilities across the range of military operations short of war. These military actions can be applied to complement any combination of the other instruments of national power and occur before, during, and after war. (JP 1-02)

military occupational specialty (MOS). Primary job of military officer or enlisted personnel.

military operations in urban terrain (MOUT). All military actions planned and conducted on a topographical complex and its adjacent natural terrain where man-made construction is the dominant feature. It includes combat-in-cities, which is that portion of MOUT involving house-to-house and street-by-street fighting in towns and cities.

military police (MP). Marine Corps and Army police forces designated by specific MOS.

maritime pre-positioning force operation. A rapid deployment and assembly of a Marine expeditionary force in a secure area using a combination of strategic airlift and forward-deployed maritime pre-positioning ships. See also Marine expeditionary force; maritime pre-positioning ships (JP 1-02 and JP 4-01.6)

nonlethal weapons (NLW). Weapons that are explicitly designed and primarily employed so as to incapacitate personnel or materiel, while minimizing fatalities, permanent injury to personnel, and undesired damage to property and the environment.

a. Unlike conventional weapons that destroy their targets through blast, penetration, and fragmentation, nonlethal weapons employ means other than gross physical destruction to prevent the target from functioning.

b. Nonlethal weapons are intended to have one, or both, of the following characteristics:

(1) They have relatively reversible effects on personnel or materiel.

(2) They affect objects differently within their area of influence. (JP 1-02)

nonlethal weapon capability set. A set of systems containing weapon systems, munitions, and protective equipment. The components are explicitly designed and primarily employed to provide additional protection to friendly personnel and to incapacitate hostile personnel or material, while minimizing fatalities or permanent injury to intended targets and minimal damage to property and the environment. They are not required to have zero probability of producing fatalities or permanent injuries but are designed and employed in a manner that significantly reduces those probabilities over the traditional employment of currently fielded military weapon systems, munitions, and equipment.

nonlethal technology (N-LT). Synonymous with NLW, but also encompassing more advanced technology that is not yet available for use by military forces.

Oleoresin Capsicum (OC). Also known as "pepper spray", OC is made completely from organic materials and is FDA-approved for sale over-the-counter in the United States. Causes severe and immediate burning sensation to mucous membranes when sprayed into face, nose, and eyes.

psychological operations (PSYOP). Planned operations to convey selected information and indicators to foreign audiences to influence their emotions, motives, objective reasoning, and ultimately the behavior of foreign governments, organizations, groups, and individuals. The purpose of psychological operations is to induce or reinforce foreign attitudes and behavior favorable to the originator's objectives. (JP 1-02)

rules of engagement (ROE). Directives issued by competent military authority that delineate the circumstances and limitations under which U.S. forces will initiate and/or continue combat engagement with other forces encountered. See also law of war. (JP 1-02)

table of organization (T/O). Total manpower authorized for specific unit, organized by billet and rank. TO& E is a U.S. Army term which encompasses T/O and T/E.

training aids and devices. Any item developed or procured with the primary intent to assist in training and learning.

weapons of mass destruction (WMD). Weapons capable of a high order of destruction and/or of being used in such a manner as to destroy large numbers of people. Weapons of mass destruction can be high explosives or nuclear, biological, chemical, and radiological weapons, but exclude the means of transporting or propelling the weapon where such means is a separable and deliverable part of the weapon. (JP 1-02)

Index

Made in the USA
Monee, IL
05 November 2026

67808356R00174